AN INTRODUCTION TO FINAN(
OPTION VALUATION
Mathematics, Stochastics and Computation

This is a lively textbook providing a solid introduction to financial option valuation for undergraduate students armed with only a working knowledge of first year calculus. Written as a series of short chapters, this self-contained treatment gives equal weight to applied mathematics, stochastics and computational algorithms, with no prior background in probability, statistics or numerical analysis required.

Detailed derivations of both the basic asset price model and the Black–Scholes equation are provided along with a presentation of appropriate computational techniques including binomial, finite differences and, in particular, variance reduction techniques for the Monte Carlo method.

Each chapter comes complete with accompanying stand-alone MATLAB code listing to illustrate a key idea. The author has made heavy use of figures and examples, and has included computations based on real stock market data. Solutions to exercises are made available at www.cambridge.org.

DES HIGHAM is a professor of mathematics at the University of Strathclyde. He has co-written two previous books, *MATLAB Guide* and *Learning LaTeX*. In 2005 he was awarded the Germund Dahlquist Prize by the Society for Industrial and Applied Mathematics for his research contributions to a broad range of problems in numerical analysis.

AN INTRODUCTION TO FINANCIAL OPTION VALUATION

Mathematics, Stochastics and Computation

DESMOND J. HIGHAM

Department of Mathematics
University of Strathclyde

CAMBRIDGE
UNIVERSITY PRESS

CAMBRIDGE UNIVERSITY PRESS
Cambridge, New York, Melbourne, Madrid, Cape Town, Singapore, São Paulo, Delhi

CAMBRIDGE UNIVERSITY PRESS
The Edinburgh Building, Cambridge CB2 2RU, UK

Published in the United States of America by Cambridge University Press, New York

www.cambridge.org
Information on this title: www.cambridge.org/9780521838849

First published 2004
Reprinted with corrections 2005, 2008

Printed in the United Kingdom at the University Press, Cambridge

Typeface Times 11/14 pt.   *System* LATEX 2_ε [TB]

A record for this publication is available from the British Library

Library of Congress in Publication data
Higham, D. J. (Desmond J.)
An introduction to financial option valuation : mathematics, stochastics, and computation /
Desmond J. Higham.
p.   cm.
Includes bibliographical references and index.
ISBN 0 521 83884 3 – ISBN 0 521 54757 1 (paperback)
1. Options (Finance) – Valuation – Mathematical models.   2. Options (Finance) – Prices – Mathematical
models.   3. Derivative securities.   I. Title.
HG6024.A3H532   2004
332.64′53 – dc22   2003069572

ISBN 978-0-521-83884-9 hardback
ISBN 978-0-521-54757-4 paperback

To my family,
Catherine, Theo, Sophie and Lucas

Contents

Illustrations

Preface

The aim of this book is to present a lively and palatable introduction to financial option valuation for undergraduate students in mathematics, statistics and related areas. Prerequisites have been kept to a minimum. The reader is assumed to have a basic competence in calculus up to the level reached by a typical first year mathematics programme. No background in probability, statistics or numerical analysis is required, although some previous exposure to material in these areas would undoubtedly make the text easier to assimilate on first reading.

The contents are presented in the form of short chapters, each of which could reasonably be covered in a one hour teaching session. The book grew out of a final year undergraduate class called *The Mathematics of Financial Derivatives* that I have taught, in collaboration with Professor Xuerong Mao, at the University of Strathclyde. The class is aimed at students taking honours degrees in Mathematics or Statistics, or joint honours degrees in various combinations of Mathematics, Statistics, Economics, Business, Accounting, Computer Science and Physics. In my view, such a class has two great selling points.

- From a student perspective, the topic is generally perceived as modern, sexy and likely to impress potential employers.
- From the perspective of a university teacher, the topic provides a focus for ideas from mathematical modelling, analysis, stochastics and numerical analysis.

There are many excellent books on option valuation. However, in preparing notes for a lecture course, I formed the opinion that there is a niche for a single, self-contained, introductory text that gives equal weight to

- applied mathematics,
- stochastics, and
- computational algorithms.

The classic applied mathematics view is provided by Wilmott, Howison and Dewynne's text (Wilmott *et al.*, 1995). My aim has been to write a book at a similar level with a less ambitious scope (only option valuation is considered), less emphasis on partial differential equations, and more attention paid to stochastic modelling and simulation.

Key features of this book are as follows.

(i) Detailed derivation and discussion of the basic lognormal asset price model.
(ii) Roughly equal weight given to binomial, finite difference and Monte Carlo methods. In particular, variance reduction techniques for Monte Carlo are treated in some detail.
(iii) Heavy use of computational examples and figures as a means of illustration.
(iv) Stand-alone MATLAB codes, with full listings and comprehensive descriptions, that implement the main algorithms. The core text can be read independently of the codes. Readers who are familiar with other programming languages or problem-solving environments should have little difficulty in translating these examples.

In a nutshell, this is the book that I wish had been available when I started to prepare lectures for the Strathclyde class.

When designing a text like this, an immediate issue is the level at which stochastic calculus is to be treated. One of the tenets of this book is that

rigorous, measure-theoretic, stochastic analysis, although beautiful, is *hard* and it is unrealistic to ask an undergraduate class to pick up such material on the fly. Monte Carlo-style simulation, on the other hand, is a relatively *simple* concept, and well-chosen computational experiments provide an excellent way to back up heuristic arguments.

Hence, the approach here is to treat stochastic calculus on a nonrigorous level and give plenty of supporting computational examples. I rely heavily on the Central Limit Theorem as a basis for heuristic arguments. This involves a deliberate compromise – convergence in distribution must be swapped for a stronger type of convergence if these arguments are to be made rigorous – but I feel that erring on the side of accessibility is reasonable, given the aims of this text.

In fact, in deriving the Black–Scholes partial differential equation, I do not make explicit reference to Itô's Lemma. I decided that a heuristic derivation of Itô's Lemma in a general setting followed by a single application of the lemma in one simple case makes less pedagogical sense than a direct 'in situ' heuristic treatment, a decision inspired by Almgren's expository article (Almgren, 2002). I hope that at least some undergraduate readers will be sufficiently motivated to follow up on the references and become exposed to the real thing.

You can get a feeling for the contents of the book by skimming through the outline bullet points that appear at the start of each chapter. Many of the later chapters can be read independently of each other, or, of course, omitted.

Exercises are given at the end of each chapter. It is my experience that active problem solving is the best learning tool, so I strongly encourage students to make use of them. I have used a starring system: one star for questions whose solution is relatively easy/short, rising to three stars for the hardest/longest questions. Brief solutions to the odd-numbered exercises are available from the book website given below. This leaves the even-numbered questions as a teaching resource. Certain questions are central to the text. I have tried to ensure that these come up in the odd-numbered list, in order to aid independent study.

A short, introductory treatment like this can only scratch the surface. Hence, each chapter concludes with a *Notes and references* section, which gives my own, necessarily biased, hints about important omissions. References can be followed up via the *References* section at the end of the book.

Scattered at the end of each chapter are a few quotes, designed to enlighten and entertain. Some of these reinforce the ideas in the text and others cast doubt on them. Mathematical option valuation is a strange business of sophisticated analysis based on simple models that have obvious flaws and perhaps do not merit such detailed scrutiny. When preparing lecture notes, I have found that authoritative, pithy quotes are a particularly powerful means to highlight some of this tension. I have an uneasy feeling that some Strathclyde students spent more time perusing the quotes than the main text, so I have aimed to make the quotes at least form a reasonable mini-summary of the contents. Most quotes relate directly to their chapter, but a few general ones have been dispersed throughout the book on the grounds that they were too good to leave out.

A website for this book has been created at www.maths.strath.ac.uk/~aas96106/ option_book.html. It includes the following.

- The MATLAB codes listed in the book.
- Outline solutions to the odd-numbered exercises.
- Links to the websites mentioned in the book.
- Colour versions of some of the figures.
- A list of corrections.
- Some extra quotes that did not make it into the book.

I am grateful to several people who have influenced this book. **Nick Higham** cast a critical eye over an early draft and made many helpful suggestions. **Vicky Henderson** checked parts of the text and patiently answered a number of questions. **Petter Wiberg** gave me access to his MATLAB files for processing stock market data. **Xuerong Mao**, through animated discussions and research collaboration, has enriched my understanding of stochastics and its role in mathematical finance. Additionally, five anonymous reviewers provided unbiased feedback. In particular, one reviewer who was not in favour of the nonrigorous approach to stochastic analysis in this book was nevertheless generous enough to provide detailed comments that allowed me to improve the final product. Finally, three years' worth of Strathclyde honours students have helped to shape my views on how to present this material to a wide audience.

MATLAB programs

I firmly believe that the best way to check your understanding of a computational algorithm is to examine, and interactively experiment with, a real program. For this reason, I have included a *Program of the Chapter* at the end of every chapter, followed by two programming exercises. Each program illustrates a key topic. They can be downloaded from the website previously mentioned.

The programs are written in MATLAB.[1] I chose this environment for a number of reasons.

- It offers excellent random number generation and graphical output facilities.
- It has powerful, built-in, high-level commands for matrix computations and statistics.
- It runs on a variety of platforms.
- It is widely available in mathematics and computer science departments and is often used as the basis for scientific computing or numerical analysis courses. Students may purchase individual copies at a modest price.

I wrote the programs with *accuracy* and *clarity* in mind, rather than efficiency or elegance. I have made quite heavy use of MATLAB's vectorization facilities, where possible working with arrays directly and eschewing unnecessary `for` loops. This tends to make the codes shorter, snappier and less daunting than alternatives that operate on individual array components. Meaningful comments have been inserted into the codes and a 'walkthrough' commentary is appended in each case. Those walkthroughs provide MATLAB information on a just-in-time basis. For a comprehensive guide to MATLAB, see (Higham and Higham, 2000).

I have not made use of any of the toolboxes that are available, at extra cost, to MATLAB users. This is because (a) the emphasis in the book is on understanding the underlying models and algorithms, not on the use of black-box packages, and (b) only a small percentage of MATLAB users will have access to toolboxes. However, those who wish to perform serious option valuation computations in MATLAB are advised to investigate the toolboxes, especially those for Finance, Statistics, Optimization and PDEs.

Readers with some experience of scientific computing in languages such as Java, C or FORTRAN should find it relatively easy to understand the codes. Those with no computing background may need to put in more effort, but should find the process rewarding.

---

[1] MATLAB is a registered trademark of The MathWorks, Inc.

MATLAB is a commercial software product produced by The Mathworks, whose homepage is at www.mathworks.com/.

Let me re-emphasize that these programs are entirely stand-alone; the book can be read without reference to them. However, I believe that they form a major element – if you understand the programs, you understand a big chunk of the material in this book.

Disclaimer of warranty

We make no warranties, express or implied, that the programs contained in this volume are free of error, or are consistent with any particular standard of merchantability, or that they will meet your requirements for any particular application. They should not be relied on for solving a problem whose incorrect solution could result in injury to a person or loss of property. If you do use the programs in such a manner, it is at your own risk. The author and publisher disclaim all liability for direct or consequential damages resulting from your use of the programs.

1

Options

1.1 What are options?

Throughout the book we use the term *asset* to describe any financial object whose value is known at present but is liable to change in the future. Typical examples are

- shares in a company,
- commodities such as gold, oil or electricity,
- currencies, for example, the value of US $100 in euros.

We will have much to say about assets in subsequent chapters, but let us get straight to the point and define an *option*.

> **Definition** A *European call option* gives its *holder* the right (but not the obligation) to purchase from the *writer* a prescribed asset for a prescribed price at a prescribed time in the future. ◇

The prescribed purchase price is known as the *exercise price* or *strike price*, and the prescribed time in the future is known as the *expiry date*.

To illustrate the idea, suppose that, today, your friend Professor Smart (the writer) writes a European call option that gives you (the holder) the right to buy 100 shares in the International Business Machines (IBM) Corporation for $1000 three months from now. After those three months have elapsed, you would then take one of two actions:

(a) if the actual value of 100 IBM shares turns out to be more than $1000 you would exercise your right to buy the shares from Professor Smart – because you could immediately sell them for a profit.

(b) if the actual value of 100 IBM shares turns out to be less than $1000 you would not exercise your right to buy the shares from Professor Smart – the deal would not be worthwhile.

Because you are not obliged to purchase the shares, you do not lose money (in case (a) you gain money and in case (b) you neither gain nor lose). Professor Smart, on the other hand, will not gain any money on the expiry date, and may lose an unlimited amount. To compensate for this imbalance, when the option is agreed (today) you would be expected to pay Professor Smart an amount of money known as the *value* of the option.

The direct opposite of a European call option is a European put option.

Definition A *European put option* gives its *holder* the right (but not the obligation) to sell to the *writer* a prescribed asset for a prescribed price at a prescribed time in the future.   ◇

The key question that we address in this book is: how much should the holder pay for the privilege of holding an option? In other words, how do we compute a fair option value?

To answer this question we have to devise a *mathematical model* for the behaviour of the asset price, come up with a precise interpretation of 'fairness' and do some analysis. These steps, which take up the next seven chapters, will lead us to the celebrated Black–Scholes formula. Looking at practical issues and more exotic options will then draw us into *computational algorithms*, which take up the bulk of the remainder of the book.

The rest of this chapter is spent on a brief review of how and why options are traded.

1.2 Why do we study options?

Options have become extremely popular; so popular that in many cases more money is invested in them than in the underlying assets. Why do they get so much attention? There are two good reasons.

(1) Options are extremely attractive to investors, both for *speculation* and for *hedging*.
(2) There is a systematic way to determine how much they are worth, and hence they can be bought and sold with some confidence.

Point (2) is the main subject of this book. To illustrate point (1), if you believe that Microsoft Corporation shares are due to increase then you may speculate by becoming the holder of a suitable call option. Typically, you can make a greater profit relative to your original payout than you would do by simply purchasing the shares. On the other hand, if you are the owner of an American company that is committed to purchasing a factory in Germany for an agreed price in euros in three months' time, then you may wish to hedge some risk by taking out an option that makes some profit in the event that the US dollar drops in value against the euro.

A further attraction is that by combining different types of option, an investor can take a position that reaps benefits from various types of asset behaviour. To understand this, it is useful to visualize options in terms of *payoff diagrams*.

We let E denote the exercise price and $S(T)$ denote the asset price at the expiry date. (Of course, $S(T)$ is not known at the time when the option is taken out.) In later chapters, $S(t)$ will be used to denote the asset price at a general time t, and T will denote the expiry date. At expiry, if $S(T) > E$ then the holder of a European call option may buy the asset for E and sell it in the market for $S(T)$, gaining an amount $S(T) - E$. On the other hand, if $E \geq S(T)$ then the holder gains nothing. Hence, we say that the *value* of the European call option at the expiry date, denoted by C, is

$$C = \max(S(T) - E, 0). \qquad (1.1)$$

Plotting $S(T)$ on the x-axis and C on the y-axis gives the payoff diagram in Figure 1.1. Consider now a European put option. If, at expiry, $E > S(T)$ then the holder may buy the asset at $S(T)$ in the market and exercise the option by selling it at E, gaining an amount $E - S(T)$. On the other hand, if $S(T) \geq E$ then the holder should do nothing. Hence, the *value* of the European put option at the expiry date, denoted by P, is

$$P = \max(E - S(T), 0). \qquad (1.2)$$

The corresponding payoff diagram is plotted in Figure 1.2. Because of their shape, the piecewise linear payoff curves in Figures 1.1 and 1.2 are sometimes referred to as (ice) *hockey sticks*.

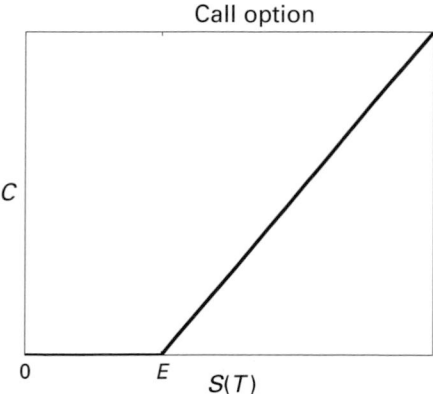

Fig. 1.1. Payoff diagram for a European call. Formula is $C = \max(S(T) - E, 0)$.

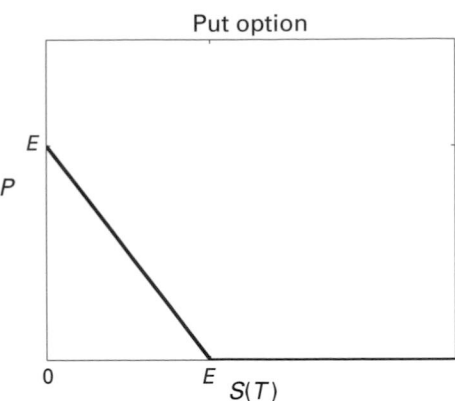

Fig. 1.2. Payoff diagram for a European put. Formula is $P = \max(E - S(T), 0)$.

Now we may plot payoff diagrams for combinations of options. For example, suppose you hold a call option and a put option on the same asset with the same expiry date and the same strike price, E. Then the overall value at expiry is the sum of $\max(S(T) - E, 0)$ and $\max(E - S(T), 0)$, which is equivalent to $|S(T) - E|$, see Exercise 1.2. This combination goes under the unfortunate name of a *bottom straddle*. The holder of a bottom straddle benefits when the asset price at expiry is *far away from the strike price* – it does not matter whether the asset finishes above or below the strike.

Another possibility is to hold a call option with exercise price E_1 and, for the same asset and expiry date, to write a call option with exercise price E_2, where $E_2 > E_1$. At the expiry date, the value of the first option is $\max(S(T) - E_1, 0)$ and the value of the second is $-\max(S(T) - E_2, 0)$. Hence, the overall value at expiry is $\max(S(T) - E_1, 0) - \max(S(T) - E_2, 0)$. The corresponding payoff diagram is plotted in Figure 1.3. This combination gives an example of a *bull spread*. We see from the figure that the holder of such a spread benefits when the asset price finishes above E_1, but gets no extra benefit if it is above E_2.

1.3 How are options traded?

Options can be traded on a number of official exchanges. The first of these, the Chicago Board Options Exchange (CBOE), started in 1973 and there are more than 50 throughout the world in 2004. Most exchanges operate through the use of *market makers*, individuals who are obliged to buy or sell options whenever asked to do so. On request, the market maker will quote a price for the option. More precisely, two prices will be quoted, the *bid* and the *ask*. The bid is the price at which the market maker will buy the option from you and the ask is the

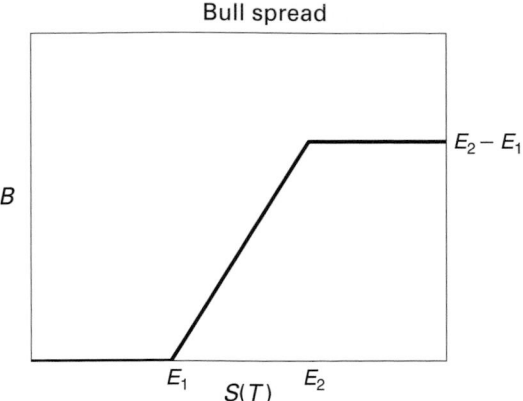

Fig. 1.3. Payoff diagram for a bull spread. Formula is $B = \max(S(T) - E_1, 0) - \max(S(T) - E_2, 0)$.

price at which the market maker will sell it to you. The bid is lower than the ask, because the market maker needs to make a living. The difference between the ask and the bid is known as the *bid–ask spread*. Typically, market makers aim to make their profits from the bid–ask spread and do not wish to speculate on the market; they seek to hedge away their risks using the type of technique that is covered in Chapters 8 and 9.

Options are also traded directly between large financial institutions – so called *over-the-counter* or OTC deals. These options often have nonstandard features that are tailored to the particular needs of the parties involved.

The *Financial Times* newspaper tabulates the prices of some options that may be traded on the London International Financial Futures & Options Exchange (LIFFE). For example, the issue from Friday, 19 September 2003 included the information

		Calls			Puts		
Option		Oct	Nov	Dec	Oct	Nov	Dec
Royal Bk Scot.	1600	67.0	92.5	109.5	29.0	49.0	62.5
(1634.0)	1700	19.5	43.5	59.0	82.0	100.0	112.5

The number 1634.0 is the closing price of The Royal Bank of Scotland's shares from the previous day. The numbers 1600 and 1700 are two exercise prices, in pence. (The *Financial Times* lists information for these exercise prices only, but the exchange offers options for many other exercise prices.) The numbers 67.0, 92.5, 109.5 are the prices of the call options with exercise price 1600 and expiry dates in

Oct, Nov and Dec, respectively (more precisely, for 18:00 on the third Wednesday of each month). Similarly, 19.5, 43.5, 59.0 are the prices of call options with exercise price 1700 for those expiry dates. The numbers 29.0, 49.0, 62.5 give the prices of put options with exercise price 1600 and expiry dates in Oct, Nov and Dec, and 82.0, 100.0, 112.5 are the corresponding put option prices for exercise price 1700. The numbers quoted lie somewhere between the bid and the ask.

The *Wall Street Journal* publishes option data in a similar form. Many providers offer electronic data access, with some basic information being available in the public domain; see Section 5.5 for some pointers.

1.4 Typical option prices

Figure 1.4 shows some prices for call and put options on IBM shares that were available on the New York Stock Exchange on 13 October 2002. Some of the data from Figure 1.4 is repeated in a slightly different format in Figure 1.5. The prevailing asset price, more precisely the price paid at the most recent trade, was 74.25, marked 'Now' in Figure 1.4. Option prices were available for a range of strike prices and expiry times. These prices relate to American, rather than European, options. Americans are introduced in Chapter 18. For the moment we note that an American call has the same value as a European call (assuming that no dividends are paid), and an American put has a higher value than a European put.

In this example, for a given expiry time, the call option price decreases as the strike price increases. This is perfectly reasonable. Increasing the strike price has a negative effect on the payoff and hence reduces the call option's worth. Similarly, the put price increases with increasing strike price. It can also be observed from the figures that, for a given strike price, both the call and the put option prices

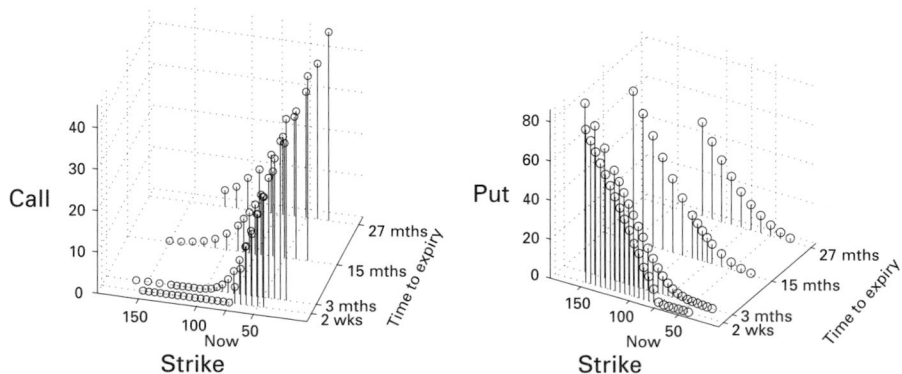

Fig. 1.4.   Market values for IBM call and put options, for a range of strike prices and times to expiry.

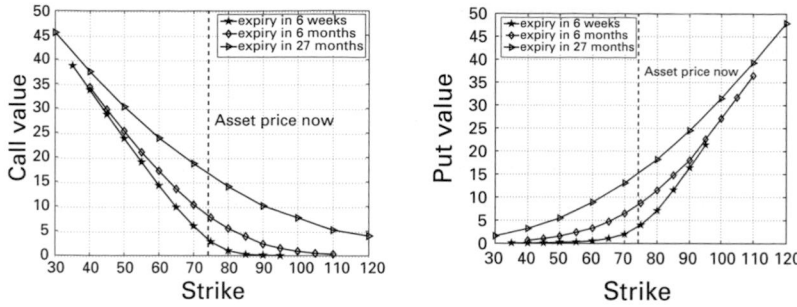

Fig. 1.5.   Market values for IBM call (left) and put (right) options, for a range of strike prices and times to expiry. This displays a subset of the data in Figure 1.4.

increase when the time to expiry increases. This behaviour is generic for European call options, as we will see in Section 2.6.

1.5 Other financial derivatives

European call and put options are the classic examples of *financial derivatives*. The term derivative indicates that their value is *derived* from the underlying asset – it has nothing to do with the mathematical meaning of a derivative. This book focuses exclusively on options. We will develop our mathematical analysis with European options in mind, and in later chapters we will introduce American and other more exotic options.

1.6 Notes and references

There are many introductory texts that explain how stock markets operate; see, for example, Dalton (2001); Walker (1991). Chapter 6 of Hull (2000) is also a good source of basic practical information about option trading, including

- what range of expiry dates and exercise prices are typically offered,
- how dividends and stock splits are dealt with, and
- how money and products actually change hands.

   Section 5.5 gives the web pages of some stock exchanges.

EXERCISES

1.1.   ⋆ Insert the word 'rise' or 'fall' to complete the following sentences:

The holder of a European call option hopes the asset price will . . .
The writer of a European call option hopes the asset price will . . .

The holder of a European put option hopes the asset price will ...
The writer of a European put option hopes the asset price will ...

1.2.   ⋆ Convince yourself that $\max(S(T) - E, 0) + \max(E - S(T), 0)$ is equivalent to $|S(T) - E|$ and draw the payoff diagram for this bottom straddle.

1.3.   ⋆⋆ Suppose that for the same asset and expiry date, you hold a European call option with exercise price E_1 and another with exercise price E_3, where $E_3 > E_1$ and also write two calls with exercise price $E_2 := (E_1 + E_3)/2$. This is an example of a *butterfly spread*.[1] Derive a formula for the value of this butterfly spread at expiry and draw the corresponding payoff diagram.

1.4.   ⋆ The holder of the bull spread with payoff diagram in Figure 1.3 would like the asset price on the expiry date to be at least as high as E_2, but, if it is, the holder does not care how much it exceeds E_2. Make similar statements about the holders of the bottom straddle in Exercise 1.2 and the butterfly spread in Exercise 1.3.

1.7 Program of Chapter 1 and walkthrough

Our first MATLAB program uses basic plotting commands to draw a bull spread payoff diagram, as shown in Figure 1.3, for particular parameters E_1 and E_2. The program is called ch01 and is stored in the file ch01.m. It is listed in Figure 1.6. The program is run by typing ch01 at the MATLAB prompt. The first three lines begin with the symbol % and hence are *comment lines*. These lines are ignored by MATLAB, they are used to provide information to humans who are reading through the code. Comment lines may be inserted anywhere, but those at the start of a code have a special property – typing help ch01 causes the information

```
CH01     Program for chapter 1
Plots a simple payoff diagram
```

to be echoed to the user. It is customary for the first comment line to begin with the name of the file in capital letters, even though the file itself has a lower case name.

The first command, clf, clears the current figure window, so that any previous graphical output is removed. The lines E1 = 2; and E2 = 4; are *assignment statements*. Variables E1 and E2 are automatically created and given those values. The semi-colon at the end of each line causes output to be suppressed. Without those semi-colons, the information

```
E1 = 2
E2 = 4
```

would be displayed on your screen. The line S = linspace(0,6,100) sets up a one-dimensional array S with 100 components, equally spaced between 0 and 6. This could be confirmed after running the program by typing S at the MATLAB prompt. The command max(S-E1,0) creates a one-dimensional array whose *i*th entry is the maximum of S(i)-E1 and 0. Note that MATLAB is happy to mix arrays and scalars, and will apply the max function in a componentwise manner. Overall

---

[1] Serve with warm toast.

the line B = max(S-E1,0) - max(S-E2,0); creates a one-dimensional array B of payoff values corresponding to S.

```
%CH01    Program for chapter 1
%
% Plots a simple payoff diagram

clf
E1 = 2;
E2 = 4;

S = linspace(0,6,100);
B = max(S-E1,0)-max(S-E2,0);
plot(S,B)
ylim([0,3])

xlabel('S')
ylabel('B')
title('Bull Spread Payoff')
grid on
```

Fig. 1.6.   Program of Chapter 1: ch01.m.

We then plot the payoff diagram with plot(S,B). By default, MATLAB chooses the range for the axes, the location of the axis tick marks, the colour and type of the line, and many other features. These may be altered with extra commands or via the menu-driven toolbars in the figure window. We have specified ylim([0,3]), which overrides the *y*-axis limits that MATLAB would otherwise choose automatically. Axis labels and a title are produced by xlabel('S'), ylabel('B') and title('Bull Spread Payoff'). The final command, grid on, causes horizontal and vertical dotted reference lines to appear in the plot. Running the program, that is, typing ch01 at the prompt, puts a picture similar to Figure 1.3 in a pop-up figure window.

Typing help linspace, help max, help plot, etc., at the command line gives more information about those functions, and MATLAB's online documentation, roused by typing doc, forms a hypertext style manual.

PROGRAMMING EXERCISES

P1.1. Use the input command to produce a variant of ch01 that allows E1 and E2 to be specified by the user.

P1.2. Create a program that plots the payoff diagram for a butterfly spread, as described in Exercise 1.3.

Quotes

Because the action is faster and the margins thinner – five percent down
will buy you a futures contract on the DAX 30 in Frankfurt,

the CAC 40 in Paris, the FTSE 100 in London, the Nikkei 225 in Tokyo,
or the Standard & Poor's 500 in New York – trading in derivatives now swamps
the markets on which they depend.

THOMAS A. BASS (Bass, 1999)

If you believe an asset will rise in price,
then you may buy a call option to capture a very large potential gain
with a small investment;
however, if your belief is wrong
then you may very easily lose your entire option investment.

ROBERT ALMGREN (Almgren, 2002)

Imagine visiting your local used-car dealer to sell him your old Ford.
He kicks the tires, points to a dent in the fender,
and offers you a hundred bucks.
Suppose the following day you are tempted to go back
and buy your Ford off the lot.
The dealer will point to the low mileage and tell you
that he can't let the car go for less than two hundred dollars.
This is the difference between the *bid*, buying price, and the *ask*, selling price.

THOMAS A. BASS (Bass, 1999)

The Pterodactyl is very rarely encountered in real trading.
The technicians may, however, wish to know
that it consists of a spread of traditional butterflies. . . .
The use of this position is not recommended unless the author needs a new car.

A. L. H. SMITH (Smith, 1986)

Recent history is replete with examples of derivatives trading gone awry.

PHILIP MCBRIDE JOHNSON (Johnson, 1999)

Winter, spring, summer or fall,
all you have to do is call. . . .

CAROL KING, *You've Got a Friend*, EMI Music Inc.

2
Option valuation preliminaries

2.1 Motivation

There are certain simple results about option valuation that can be deduced from first principles, using elementary mathematics. This chapter derives such results. To do this we introduce two key concepts: discounting for interest and the no arbitrage principle. The results that we derive do not require us to make any assumptions about the behaviour of the underlying asset, nor do they use any probability theory.

2.2 Interest rates

Suppose we have some money in a risk-free savings account. If this investment grows according to a *continuously compounded interest rate, r*, then its value increases by a factor e^{rt} over a time length t. In other words, an amount D_0 at time zero is worth

$$D(t) = e^{rt} D_0 \tag{2.1}$$

at time t. To be specific, we will use r to denote the *annual* rate, so that time is measured in years. Typical values of r lie between 0.01 and 0.1 (1% and 10% interest rates). It is not important in this book whether $D(t)$ is measured in dollars, euros, or any other currency.

11

Throughout the book we make the standing assumption that a fixed interest rate r prevails whenever cash is lent or borrowed; the same rate r applies at all times and whatever amount of cash is involved. This assumption is, of course, only an approximation to reality – interest rates change over time and typically depend on the size of the investment.

An immediate consequence of our interest rate assumption is that if somebody were to make you the offer of

(a) \$100 immediately (time $t = 0$), or
(b) \$100e^{rt} at time t

then you would regard both offers as being of equal value. (In case (a) you could invest the money to obtain \$100e^{rt} at time t. In case (b) you could borrow \$100 immediately and repay the loan at time t.)

Similarly, a deal that is guaranteed to produce exactly \$100 at time t is worth exactly \$100e^{-rt} at time zero. Transferring from \$100 to \$100e^{-rt} in this way is called *discounting for interest* or *discounting for inflation*, and it is a concept that we will use frequently.

2.3 Short selling

We use the term *portfolio* to describe a combination of

  (i) assets,
 (ii) options, and
(iii) cash (invested in a bank).

Moreover, we will assume that it is possible to hold negative amounts of each. A negative amount of cash has the obvious interpretation that cash has been borrowed rather than invested in the bank. Owning a negative amount of asset or option might not seem so reasonable. However, in many cases this is possible through the practice of *short selling*, which means selling an item that is not owned with the intention of buying it back at a later date. In practice, to short sell an item you must first borrow it from somebody who owns it, and give it back later. We will assume that this is always possible, at no cost, and that the short seller is free to choose when to buy back and return the item.

To illustrate the idea, let $S(t)$ denote the value of an asset at time t. If we short sell an asset at time t_1 and buy it back at time t_2, then we have

(a) gained an amount $S(t_1)$ at time $t = t_1$ from the short sale,
(b) paid out an amount $S(t_2)$ at time $t = t_2$ from the buy back.

Having invested the initial gain, the overall profit/loss at time $t = t_2$ is therefore $e^{r(t_2-t_1)}S(t_1) - S(t_2)$.

2.4 Arbitrage

One of the key principles on which option valuation theory rests is *no arbitrage*. This may be summarized as follows.

There is never an opportunity to make a risk-free profit that gives a greater return than that provided by the interest from a bank deposit.

Note that this assumption applies only to **risk-free** profit, it is not relevant to portfolios that 'have a good chance' of making a greater return than a bank deposit.

To justify the no arbitrage assumption, suppose it were possible to put together a portfolio that gave a guaranteed improvement on the bank's interest rate. Sensible investors would simply borrow money from the bank and spend it on the portfolio, thereby locking in to a guaranteed risk-free profit. The forces of supply and demand would then cause the yield from the portfolio to drop, or the interest rate to increase, or both, until parity was restored. Further justification for this assumption is provided by the existence of *arbitrageurs* who scour the markets seeking to exploit any opportunities for risk-free profits beyond the interest rate level.

2.5 Put–call parity

There is a delightfully simple argument that defines a relationship between the value C of a European call option and the value P of a European put option, with the same strike price E and expiry date T. (In this section and the next, value is taken by default to mean value at time $t = 0$.) Consider two portfolios

> π_A: one call option plus Ee^{-rT} cash (invested in a bank),
> π_B: one put option plus one unit of the asset.

At the expiry date, the portfolio π_A is worth $\max(S(T) - E, 0) + E$, which is $\max(S(T), E)$. The portfolio π_B is worth on expiry $\max(E - S(T), 0) + S(T)$, which also reduces to $\max(S(T), E)$. Common sense dictates that since the two portfolios always give the same payoff, they must have the same value at time zero, so

$$C + Ee^{-rT} = P + S. \tag{2.2}$$

This relationship, which connects the values of the call and put, is called *put–call parity*. Note that (2.2) was derived without any assumptions about the behaviour of the asset. Because of put–call parity, if we can work out a procedure for valuing a European call option, we automatically get a procedure for valuing a European put option, and vice versa.

The argument behind (2.2) can be made more precise via the no arbitrage principle. If π_A were worth more than π_B at time 0 then it would be possible to sell π_A (that is, sell the call option and borrow the cash) and buy π_B (that is, buy one put option and one share). This brings us an instantaneous profit of $\pi_A - \pi_B$ (since we are sure that the payoff from π_B exactly compensates for that of π_A at expiry). Such instantaneous profit clearly violates the no arbitrage principle. A similar argument applies if π_B is worth more than π_A at time zero.

2.6 Upper and lower bounds on option values

Similar arguments to those above can be used to obtain simple upper and lower bounds on the values C and P of European call and put options.

To study the call option, consider two portfolios:

π_A: one call option plus Ee^{-rT} cash (invested in a bank),
$\widehat{\pi}_B$: one unit of asset.

We saw above that π_A has payoff $\max(S(T), E)$ at expiry. The portfolio $\widehat{\pi}_B$ has payoff $S(T)$, which is never greater than the payoff for π_A. Common sense (or, more formally, the no arbitrage assumption – Exercise 2.3) dictates that $\widehat{\pi}_B$ must therefore have a time-zero value that is no greater than that for π_A. This means $S \leq C + Ee^{-rT}$, or

$$C \geq S - Ee^{-rT}. \tag{2.3}$$

Since the call option cannot have a negative value, we may strengthen this to

$$C \geq \max(S - Ee^{-rT}, 0). \tag{2.4}$$

On the other hand, since the strike price E is always ≥ 0, the call option can never be worth more than the underlying asset, so

$$C \leq S. \tag{2.5}$$

Figure 2.1 illustrates the bounds (2.4) and (2.5).

The corresponding upper and lower bounds for P,

$$P \geq \max(Ee^{-rT} - S, 0) \qquad \text{and} \qquad P \leq Ee^{-rT}, \tag{2.6}$$

can be derived either by a similar argument, or via (2.4)–(2.5) and put–call parity (2.2), see Exercise 2.4.

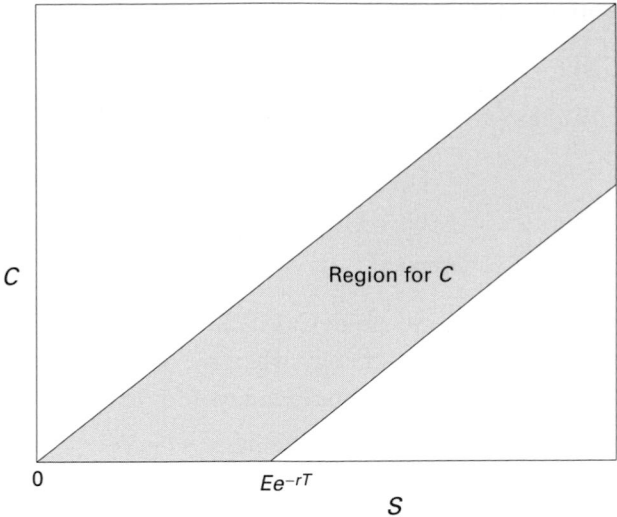

Fig. 2.1. Upper and lower bounds (2.4) and (2.5) for European call option. Here, the x-axis is S, the asset price at time zero, and the y-axis is C, option value at time zero. Option value C must lie in the shaded region.

A final result that we can prove from first principles is as follows.

The time-zero European call option value, C, is nondecreasing as a function of the expiry date T.

To see this, consider European call options with expiry dates T_1 and T_2, with $T_2 > T_1$, having the same strike price, E. We will show that the holder of the T_2 option can guarantee to get a payoff at least as big as $e^{r(T_2-T_1)} \max(S(T_1) - E, 0)$. Suppose the T_2 option holder takes the following action at time $t = T_1$.

Case 1: If $S(T_1) \leq E$ do nothing. (The T_1 option has zero payoff, so the T_2 option payoff will be no worse.)

Case 2: If $S(T_1) > E$ then short sell one unit of the asset at time $t = T_1$, invest the money, and buy back the asset at $t = T_2$. (Intuitively, the T_1 option produces a positive payoff. In order to match it, the T_2 holder guards against future decrease in the asset price by taking out an investment that gains if the asset falls.)

In Case 1 it is trivially true that the T_2 option holder makes an overall profit of at least $e^{r(T_2-T_1)} \max(S(T_1) - E, 0) = 0$. In Case 2, the T_2 option holder has a payoff at time $t = T_2$ made up of

(a) $\max(S(T_2) - E, 0)$ from the original T_2 option, plus
(b) $e^{r(T_2-T_1)} S(T_1)$ from investing the proceeds of short selling the asset at time $t = T_1$, plus
(c) $-S(T_2)$ from covering the short sale.

The overall payoff at $t = T_2$ is thus

$$
\max(S(T_2) - E, 0) + e^{r(T_2 - T_1)} S(T_1) - S(T_2)
$$
$$
= \max \left(e^{r(T_2 - T_1)} S(T_1) - E, \, e^{r(T_2 - T_1)} S(T_1) - S(T_2) \right)
$$
$$
\geq e^{r(T_2 - T_1)} S(T_1) - E
$$
$$
\geq e^{r(T_2 - T_1)} \left(S(T_1) - E \right)
$$
$$
= e^{r(T_2 - T_1)} \max(S(T_1) - E, 0).
$$

(The last line follows because we are in Case 2.)

We have shown that there is a strategy whereby, after discounting for interest, the T_2 option holder can guarantee to have a payoff at least as great as that of the T_1 option holder. Hence, by the no arbitrage principle, the T_2 option must have a value at least as great as that of the T_1 option.

It is perhaps surprising that there is no such simple result for European put options; see Exercise 10.7 in Chapter 10.

This chapter has given an indication of some simple results that can be derived from first principles. To proceed further we need to make assumptions about the behaviour of the underlying asset, which leads us immediately into the realms of probability and random variables.

2.7 Notes and references

Further details about arbitraging and short selling can be found, for example, in (Hull, 2000).

EXERCISES

2.1. ⋆ *Compound interest* works as follows. An investment D_0 at time zero when compounded m times up to time t at rate r_c becomes worth

$$
D(t) = \left(1 + \frac{r_c t}{m} \right)^m D_0.
$$

Show that, for a given m, the compound interest rate r_c that produces the same amount as the continuously compounded value $e^{rt} D_0$ satisfies $r_c = m(e^{rt/m} - 1)/t$. Use the approximation $e^x \approx 1 + x$ for small x to show that $r_c \approx r$ when m is large. (Note that in this book we always work with continuously compounded interest.)

2.2. ⋆⋆ The continuously compounded interest rate formula can be derived by

(a) splitting the time interval $[0, t]$ into subintervals $[0, \delta t]$, $[\delta t, 2\delta t]$, ..., $[(L - 1) \delta t, L\delta t]$, where $\delta t = t/L$, and

(b) assuming that the value of the investment increases by a relative amount proportional to $r\delta t$ over each subinterval.

Letting $t_i = i\delta t$, this means

$$D(t_{i+1}) = (1 + r\delta t)D(t_i), \tag{2.7}$$

and hence

$$D(t = t_L) = (1 + r\delta t)^L D_0.$$

By writing this as $D(t) = e^{L \log(1+rt/L)} D_0$ and using $\log(1 + \epsilon) = \epsilon + O(\epsilon^2)$ as $\epsilon \to 0$, show that this model reproduces the formula (2.1) in the limit $L \to \infty$ (i.e. $\delta t \to 0$). Show that the models

$$D(t_{i+1}) = \left(1 + r\sqrt{\delta t}\right) D(t_i) \tag{2.8}$$

and

$$D(t_{i+1}) = \left(1 + r\,(\delta t)^{\frac{3}{2}}\right) D(t_i) \tag{2.9}$$

are not consistent with continuous compounding in the limit $L \to \infty$.

2.3. ★★ Give an argument based on the no arbitrage assumption that justifies (2.3).

2.4. ★★ Establish (2.6) (a) by setting up suitable portfolios and applying the arguments used to get (2.4)–(2.5), and (b) separately, by using (2.4)–(2.5) plus put–call parity (2.2).

2.5. ★★★ Show that a butterfly spread with exactly the same payoff as that in Exercise 1.3 can be obtained using only a combination of European put options. Use put–call parity (2.2) to confirm that the two spreads have the same set-up cost.

2.6. ★★★ A *forward contract*, which is similar to a *futures contract*, operates as follows. Now, at time $t = 0$, Party A agrees to purchase an asset from Party B at a specified delivery time $t = T$ for a specified price F. (Note that Party A is committed to the future purchase – by contrast, with a European call option the holder has the right, but not the obligation, to buy at the prescribed price.) Appealing to the no arbitrage assumption, show that a fair value for F is $S(0)e^{rT}$.

2.8 Program of Chapter 2 and walkthrough

The program ch02, which illustrates the connection between compound and continuous interest covered in Exercise 2.1, makes use of MATLAB's for loop construction. It is listed in Figure 2.2. After clearing the figure window and initializing dzero, r, T and m, we use a for loop to do the main computation. The syntax for i = 1:m ...end causes the enclosed statements to be executed m times, first with i=1, then with i=2, and so on up to i=m. It follows from Exercise 2.1 that d(i) is the value of the investment after i months, corresponding to time tval(i).

```
%CH02    Program for Chapter 2
%
% Illustrates compound interest

clf
dzero = 5;
r = 0.15;        % Compound interest rate
T = 5;           % 5 year period
m = 60;          % 60 months

for i = 1:m      % let i months elapse
   tval(i) = i/12; % time in years
   d(i) = dzero*(1+r*tval(i)/i)^i; % compound i times
end

trange = [0,tval];
drange = [dzero,d];

plot(trange,drange,'r*')
hold on
tcts = linspace(0,T,100);
dcts = dzero*exp(r*tcts);
plot(tcts,dcts,'b-')
grid on
xlabel('t')
ylabel('D(t)')
title('Compound versus Continuous Interest')
legend('Compound','Continuous',2)
```

Fig. 2.2.   Program of Chapter 2: ch02.m.

The line `trange = [0,tval]` creates a new one-dimensional array whose first entry is 0, second entry is `tval(1)`, third entry is `tval(2)`, etc. Similarly `drange` has entries `dzero`, `d(1)`, `d(2)`,..., `d(m)`. This is done so that the initial values are included in the plot. The `legend` function produces the annotation seen at the top right-hand corner of the picture. The remainder of the program uses commands explained in Chapter 1. The program produces the picture shown in Figure 2.3.

PROGRAMMING EXERCISES

P2.1. Adapt ch02 to show that the models (2.8) and (2.9) are not sensible.

P2.2. Use the command `fill` to create a picture similar to Figure 2.1.

Quotes

The *principle* of no-arbitrage pricing is obvious,
but its application leads to many subtle and unanticipated pricing relationships.

JOHN H. COCHRANE (Cochrane, 2001)

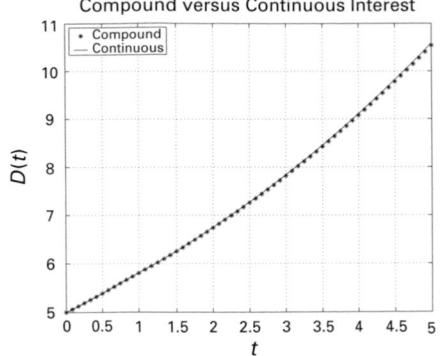

Fig. 2.3.   Figure produced by ch02.

In need of a euphemism for what we did with other people's money,
we called it 'arbitrage',
which was just plain obfuscation.

MICHAEL LEWIS (Lewis, 1989)

In practical terms, those who go short sell a security they have borrowed.
They must return the security later – by which time, they believe,
the price will have declined.
The principle of buying cheap and selling dear still holds.
Short sellers merely reverse the order: sell dear, *then* buy cheap.

ROGER LOWENSTEIN (Lowenstein, 2001)

Others are contemplating a bet known as the O'Hare straddle.
You max out your credit line,
make a stab at guessing tomorrow's opening prices,
and flee to O'Hare Airport.
You wake up tomorrow in Rio, either a bankrupt fugitive
or a lucky millionaire.

THOMAS A. BASS (Bass, 1999)

3

Random variables

3.1 Motivation

The mathematical ideas that we develop in this book are going to involve *random variables*. In this chapter we give a very brief introduction to the main ideas that are needed. If this material is completely new to you, then you may need to refer back to this chapter as you progress through the book.

3.2 Random variables, probability and mean

If we roll a fair dice, each of the six possible outcomes $1, 2, \ldots, 6$ is equally likely. So we say that each outcome has probability $1/6$. We can generalize this idea to the case of a *discrete random variable* X that takes values from a finite set of numbers $\{x_1, x_2, \ldots, x_m\}$. Associated with the random variable X are a set of *probabilities* $\{p_1, p_2, \ldots, p_m\}$ such that x_i occurs with probability p_i. We write $\mathbb{P}(X = x_i)$ to mean 'the probability that $X = x_i$'. For this to make sense we require

- $p_i \geq 0, \quad$ for all $i \qquad$ (negative probabilities not allowed),
- $\sum_{i=1}^{m} p_i = 1 \qquad$ (probabilities add up to 1).

The *mean*, or *expected value*, of a discrete random variable X, denoted by $\mathbb{E}(X)$, is defined by

$$\mathbb{E}(X) := \sum_{i=1}^{m} x_i \, p_i. \tag{3.1}$$

Note that for the dice example above we have

$$\mathbb{E}(X) = \frac{1}{6}1 + \frac{1}{6}2 + \cdots + \frac{1}{6}6 = \frac{6+1}{2},$$

which is intuitively reasonable.

Example A random variable X that takes the value 1 with probability p (where $0 \le p \le 1$) and takes the value 0 with probability $1 - p$ is called a *Bernoulli random variable with parameter* p. Here, $m = 2$, $x_1 = 1$, $x_2 = 0$, $p_1 = p$ and $p_2 = 1 - p$, in the notation above. For such a random variable we have

$$\mathbb{E}(X) = 1p + 0(1 - p) = p. \tag{3.2}$$

$\diamond$

A *continuous random variable* may take any value in $\mathbb{R}$. In this book, continuous random variables are characterized by their density functions. If X is a continuous random variable then we assume that there is a real-valued *density function* f such that the probability of $a \le X \le b$ is found by integrating $f(x)$ from $x = a$ to $x = b$; that is,

$$\mathbb{P}(a \le X \le b) = \int_a^b f(x)dx. \tag{3.3}$$

Here, $\mathbb{P}(a \le X \le b)$ means 'the probability that $a \le X \le b$'. For this to make sense we require

- $f(x) \ge 0,$ for all x (negative probabilities not allowed),
- $\int_{-\infty}^{\infty} f(x)dx = 1$ (density integrates to 1).

The *mean*, or *expected value*, of a continuous random variable X, denoted $\mathbb{E}(X)$, is defined by

$$\mathbb{E}(X) := \int_{-\infty}^{\infty} xf(x)dx. \tag{3.4}$$

Note that in some cases this infinite integral does not exist. In this book, whenever we write $\mathbb{E}$ we are implicitly assuming that the integral exists.

Example A random variable X with density function

$$f(x) = \begin{cases} (\beta - \alpha)^{-1}, & \text{for } \alpha < x < \beta, \\ 0 & \text{otherwise}, \end{cases} \tag{3.5}$$

is said to have a *uniform distribution* over (α, β). We write $X \sim \mathsf{U}(\alpha, \beta)$. Loosely, X only takes values between α and β and is equally likely to take any such value. More precisely, given values x_1 and x_2 with $\alpha < x_1 < x_2 < \beta$, the probability that X takes a value in the interval $[x_1, x_2]$ is given by the relative size of the interval: $(x_2 - x_1)/(\beta - \alpha)$. Exercise 3.1 asks you to confirm this. If $X \sim U(\alpha, \beta)$ then X has mean given by

$$\mathbb{E}(X) = \int_{-\infty}^{\infty} x f(x) dx = \frac{1}{\beta - \alpha} \int_{\alpha}^{\beta} x dx = \frac{1}{\beta - \alpha} \left[\frac{x^2}{2} \right]_{\alpha}^{\beta} = \frac{\alpha + \beta}{2}.$$

$\diamond$

Generally, if X and Y are random variables, then we may create new random variables by combining them. So, for example, $X + Y$, $X^2 + \sin(Y)$ and $e^{\sqrt{X+Y}}$ are also random variables.

Two fundamental identities that apply for any random variables X and Y are

$$\mathbb{E}(X + Y) = \mathbb{E}(X) + \mathbb{E}(Y), \tag{3.6}$$

$$\mathbb{E}(\alpha X) = \alpha \mathbb{E}(X), \qquad \text{for } \alpha \in \mathbb{R}. \tag{3.7}$$

In words: the mean of the sum is the sum of the means, and the mean scales linearly. The following result will also prove to be very useful. If we apply a function h to a continuous random variable X then the mean of the random variable $h(X)$ is given by

$$\mathbb{E}(h(X)) = \int_{-\infty}^{\infty} h(x) f(x) dx. \tag{3.8}$$

3.3 Independence

If we say that the two random variables X and Y are *independent*, then this has an intuitively reasonable interpretation – the value taken by X does not depend on the value taken by Y, and vice versa. To state the classical, formal definition of independence requires more background theory than we have given here, but an equivalent condition is

$$\mathbb{E}(g(X)h(Y)) = \mathbb{E}(g(X))\mathbb{E}(h(Y)), \qquad \text{for all } g, h : \mathbb{R} \to \mathbb{R}.$$

In particular, taking g and h to be the identity function, we have

$$X \text{ and } Y \text{ independent} \implies \mathbb{E}(XY) = \mathbb{E}(X)\mathbb{E}(Y). \tag{3.9}$$

Note that $\mathbb{E}(XY) = \mathbb{E}(X)\mathbb{E}(Y)$ does not hold, in general, when X and Y are not independent. For example, taking X as in Exercise 3.4 and $Y = X$ we have $\mathbb{E}(X^2) \neq (\mathbb{E}(X))^2$.

We will sometimes encounter *sequences* of random variables that are *independent and identically distributed*, abbreviated to *i.i.d.* Saying that $X_1, X_2, X_3, \ldots$ are i.i.d. means that

(i) in the discrete case the X_i have the same possible values $\{x_1, x_2, \ldots, x_m\}$ and probabilities $\{p_1, p_2, \ldots, p_m\}$, and in the continuous case the X_i have the same density function $f(x)$, and

(ii) being told the values of any subset of the X_is tells us nothing about the values of the remaining X_is.

In particular, if $X_1, X_2, X_3, \ldots$ are i.i.d. then they are pairwise independent and hence $\mathbb{E}(X_i X_j) = \mathbb{E}(X_i)\mathbb{E}(X_j)$, for $i \neq j$.

3.4 Variance

Having defined the mean of discrete and continuous random variables in (3.1) and (3.4), we may define the *variance* as

$$\text{var}(X) := \mathbb{E}((X - \mathbb{E}(X))^2). \tag{3.10}$$

Loosely, the mean tells you the 'typical' or 'average' value and the variance gives you the amount of 'variation' around this value.

The variance has the equivalent definition

$$\text{var}(X) := \mathbb{E}(X^2) - (\mathbb{E}(X))^2; \tag{3.11}$$

see Exercise 3.3. That exercise also asks you to confirm the scaling property

$$\text{var}(\alpha X) = \alpha^2 \text{var}(X), \qquad \text{for } \alpha \in \mathbb{R}. \tag{3.12}$$

The *standard deviation*, which we denote by std, is simply the square root of the variance; that is

$$\text{std}(X) := \sqrt{\text{var}(X)}. \tag{3.13}$$

Example Suppose X is a Bernoulli random variable with parameter p, as introduced above. Then $(X - \mathbb{E}(X))^2$ takes the value $(1 - p)^2$ with probability p and p^2 with probability $1 - p$. Hence, using (3.10),

$$\text{var}(X) = \mathbb{E}((X - \mathbb{E}(X))^2) = (1 - p)^2 p + p^2(1 - p) = p - p^2. \tag{3.14}$$

It follows that taking $p = \frac{1}{2}$ maximizes the variance.   ◇

Example For $X \sim \mathsf{U}(\alpha, \beta)$ we have $\mathbb{E}(X^2) = (\alpha^2 + \alpha\beta + \beta^2)/3$ and hence, from (3.11), $\text{var}(X) = (\beta - \alpha)^2/12$, see Exercise 3.5. So, if $Y_1 \sim \mathsf{U}(-1, 1)$ and $Y_2 \sim \mathsf{U}(-2, 2)$, then Y_1 and Y_2 have the same mean, but Y_2 has a bigger variance, as we would expect.   ◇

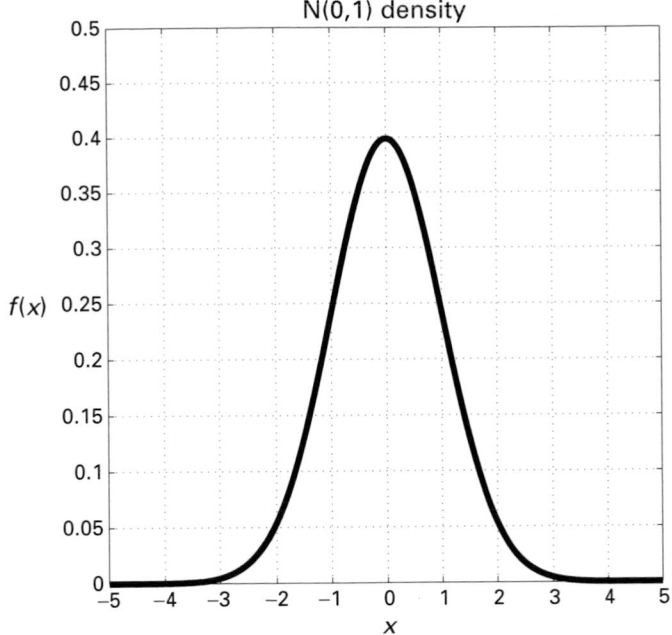

Fig. 3.1.   Density function (3.15) for an N(0, 1) random variable.

3.5 Normal distribution

One particular type of random variable turns out to be by far the most important for our purposes (and indeed for most purposes). If X is a continuous random variable with density function

$$f(x) = \frac{1}{\sqrt{2\pi}} e^{-\frac{x^2}{2}}, \tag{3.15}$$

then we say that X has the *standard normal distribution* and we write $X \sim N(0, 1)$. Here N stands for normal, 0 is the mean and 1 is the variance; so for this X we have $\mathbb{E}(X) = 0$ and $\text{var}(X) = 1$, see Exercise 3.7. Plotting the density f in (3.15) reveals the familiar *bell-shaped curve*; see Figure 3.1.

More generally, a $N(\mu, \sigma^2)$ random variable, which is characterized by the density function

$$f(x) = \frac{1}{\sqrt{2\pi\sigma^2}} e^{-\frac{(x-\mu)^2}{2\sigma^2}}, \tag{3.16}$$

has mean μ and variance σ^2; see Exercise 3.8. Figure 3.2 plots density functions for various μ and σ. The curves are symmetric about $x = \mu$. Increasing the variance σ^2 causes the density to flatten out – making extreme values more likely.

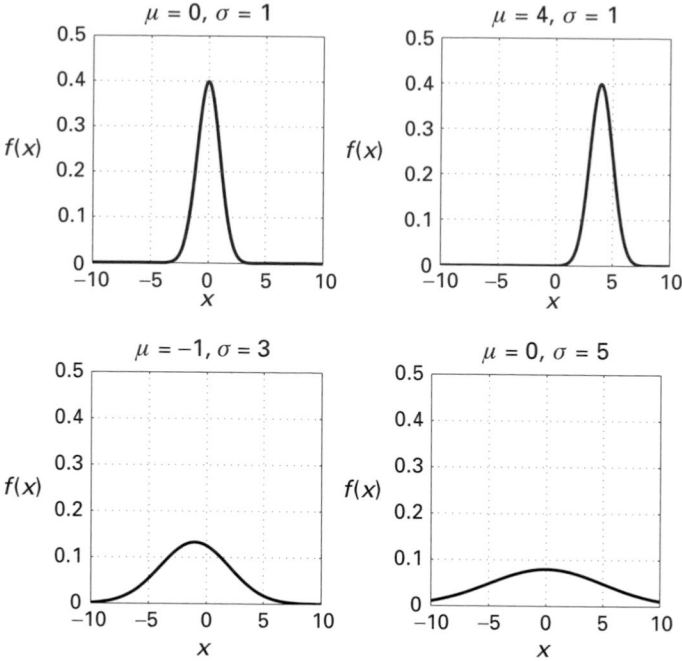

Fig. 3.2. Density functions for various $N(\mu, \sigma^2)$ random variables.

Given a density function $f(x)$ for a continuous random variable X, we may define the *distribution function* $F(x) := \mathbb{P}(X \le x)$, or, equivalently,

$$F(x) := \int_{-\infty}^{x} f(s)\, ds. \tag{3.17}$$

In words, $F(x)$ is the area under the density curve to the left of x. The distribution function for a standard normal random variable turns out to play a central role in this book, so we will denote it by $N(x)$:

$$N(x) := \frac{1}{\sqrt{2\pi}} \int_{-\infty}^{x} e^{-\frac{s^2}{2}}\, ds. \tag{3.18}$$

Figure 3.3 gives a plot of $N(x)$.

Some useful properties of normal random variables are:

(i) If $X \sim N(\mu, \sigma^2)$ then $(X - \mu)/\sigma \sim N(0, 1)$.
(ii) If $Y \sim N(0, 1)$ then $\sigma Y + \mu \sim N(\mu, \sigma^2)$.
(iii) If $X_1 \sim N(\mu_1, \sigma_1^2)$, $X_2 \sim N(\mu_2, \sigma_2^2)$ and X_1 and X_2 are independent, then $X_1 + X_2 \sim N(\mu_1 + \mu_2, \sigma_1^2 + \sigma_2^2)$.

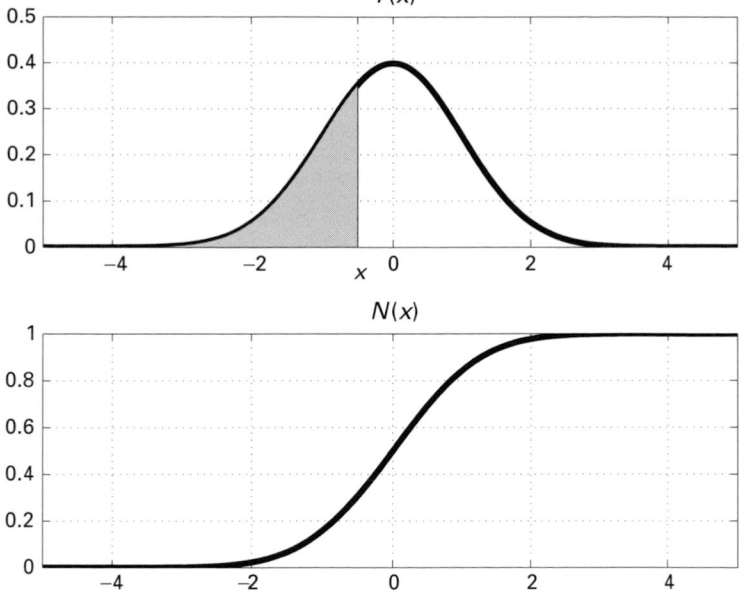

Fig. 3.3. Upper picture: N(0, 1) density. Lower picture: the distribution function
$N(x)$ – for each x this is the area of the shaded region in the upper picture.

3.6 Central Limit Theorem

A fundamental, beautiful and far-reaching result in probability theory says that the
sum of a large number of i.i.d. random variables will be approximately normal.
This is the *Central Limit Theorem*. To be more precise, let $X_1, X_2, X_3, \ldots$ be a
sequence of i.i.d. random variables, each with mean μ and variance σ^2, and let

$$S_n := \sum_{i=1}^{n} X_i.$$

The Central Limit Theorem says that for large n, S_n behaves like an $N(n\mu, n\sigma^2)$
random variable. More precisely, $(S_n - n\mu)/(\sigma\sqrt{n})$ is approximately $N(0, 1)$ in
the sense that for any x we have

$$\mathbb{P}\left(\frac{S_n - n\mu}{\sigma\sqrt{n}} \leq x\right) \to N(x), \qquad \text{as } n \to \infty. \tag{3.19}$$

The result (3.19) involves *convergence in distribution*. It says that the distribution function for $(S_n - n\mu)/(\sigma\sqrt{n})$ converges pointwise to $N(x)$. There are many
other, distinct senses in which a sequence of random variables may exhibit some
sort of limiting behaviour, but none of them will be discussed in this book. So
whenever we argue that a sequence of random variables is 'close to some random

variable X', we implicitly mean close in this distributional sense. We will be using the Central Limit Theorem as a means to derive heuristically a number of stochastic expressions. Justifying these derivations rigorously would require us to introduce stronger concepts of convergence and set up some technical machinery. To keep the book as accessible as possible, we have chosen to avoid this route. Fortunately, the Central Limit Theorem does not lead us astray.

An awareness of the Central Limit Theorem has led many scientists to make the following logical step: real-life systems are subject to a range of external influences that can be reasonably approximated by i.i.d. random variables and hence the overall effect can be reasonably modelled by a single normal random variable with an appropriate mean and variance. This is why normal random variables are ubiquitous in stochastic modelling. With this in mind, it should come as no surprise that normal random variables will play a leading role when we tackle the problem of modelling assets and valuing financial options.

3.7 Notes and references

The purpose of this chapter was to equip you with the minimum amount of material on random variables and probability that is needed in the rest of the book. As such, it has left a vast amount unsaid. There are many good introductory books on the subject. A popular choice is (Grimmett and Welsh, 1986), which leads on to the more advanced text (Grimmett and Stirzaker, 2001).

Lighter reading is provided by two highly accessible texts of a more informal nature, (Isaac, 1995) and (Nahin, 2000).

A comprehensive, introductory text that may be freely downloaded from the WWW is (Grinstead and Snell, 1997). This book, and many other resources, can be found via *The Probability Web* at http://mathcs.carleton.edu/probweb/probweb.html.

To study probability with complete rigour requires the use of measure theory. Accessible routes into this area are offered by (Capiński and Kopp, 1999) and (Rosenthal, 2000).

EXERCISES

3.1. ⋆ Suppose $X \sim \mathsf{U}(\alpha, \beta)$. Show that for an interval $[x_1, x_2]$ in (α, β) we have

$$\mathbb{P}(x_1 \leq X \leq x_2) = \frac{x_2 - x_1}{\beta - \alpha}.$$

3.2. ⋆⋆ Show that (3.7) holds for a discrete random variable. Now suppose that X is a continuous random variable with density function f. Recall that the density function is characterized by (3.3). What is the density function of αX, for $\alpha \in \mathbb{R}$? Show that (3.7) holds.

3.3. ⋆⋆ Using (3.6) and (3.7) show that (3.10) and (3.11) are equivalent and establish (3.12).

3.4. ⋆⋆ A continuous random variable X with density function

$$f(x) = \begin{cases} \lambda e^{-\lambda x}, & \text{for } x > 0, \\ 0, & \text{for } x \leq 0, \end{cases}$$

where $\lambda > 0$, is said to have the *exponential distribution with parameter* λ. Show that in this case $\mathbb{E}(X) = 1/\lambda$. Show also that $\mathbb{E}(X^2) = 2/\lambda^2$ and hence find an expression for $\text{var}(X)$.

3.5. ⋆⋆ Show that if $X \sim \mathsf{U}(\alpha, \beta)$ then $\mathbb{E}(X^2) = (\alpha^2 + \alpha\beta + \beta^2)/3$ and hence $\text{var}(X) = (\beta - \alpha)^2/12$.

3.6. ⋆ Let X and Y be independent random variables and let $\alpha \in \mathbb{R}$ be a constant. Show that $\text{var}(X + Y) = \text{var}(X) + \text{var}(Y)$ and $\text{var}(\alpha + X) = \text{var}(X)$.

3.7. ⋆⋆⋆ Suppose that $X \sim \mathsf{N}(0, 1)$. Verify that $\mathbb{E}(X) = 0$. From (3.8), the *second moment* of X, $\mathbb{E}(X^2)$, satisfies

$$\mathbb{E}(X^2) = \frac{1}{\sqrt{2\pi}} \int_{-\infty}^{\infty} x^2 e^{-x^2/2} dx.$$

Using integration by parts, show that $\mathbb{E}(X^2) = 1$, and hence that $\text{var}(X) = 1$. From (3.8) again, for any integer $p > 0$ the *pth moment* of X, $\mathbb{E}(X^p)$, satisfies

$$\mathbb{E}(X^p) = \frac{1}{\sqrt{2\pi}} \int_{-\infty}^{\infty} x^p e^{-x^2/2} dx.$$

Show that $\mathbb{E}(X^3) = 0$ and $\mathbb{E}(X^4) = 3$, and find a general expression for $\mathbb{E}(X^p)$. (Note: you may use without proof the fact that $\int_{-\infty}^{\infty} e^{-x^2/2} dx = \sqrt{2\pi}$.)

3.8. ⋆⋆ From the definition (3.16) of its density function, verify that an $\mathsf{N}(\mu, \sigma^2)$ random variable has mean μ and variance σ^2.

3.9. ⋆⋆ Show that $N(x)$ in (3.18) satisfies $N(\alpha) + N(-\alpha) = 1$.

3.8 Program of Chapter 3 and walkthrough

As an alternative to the four separate plots in Figure 3.2, ch03, listed in Figure 3.4, produces a three-dimensional plot of the $\mathsf{N}(0, \sigma^2)$ density function as σ varies. The new commands introduced are meshgrid and waterfall. We look at σ values between 1 and 5 in steps of dsig $= 0.25$ and plot

```
%CH03    Program for Chapter 3
%
% Illustrates Normal distribution

clf

dsig = 0.25;
dx = 0.5;
mu = 0;
[X,SIGMA] = meshgrid(-10:dx:10,1:dsig:5);
Z = exp(-(X-mu).^2./(2*SIGMA.^2))./sqrt(2*pi*SIGMA.^2);
waterfall(X,SIGMA,Z)
xlabel('x')
ylabel('\sigma')
zlabel('f(x)')
title('N(0,\sigma) density for various \sigma')
```

Fig. 3.4.   Program of Chapter 3: `ch03.m`.

the density function for x between -10 and 10 in steps of `dx = 0.5`. The line

$$\texttt{[X,SIGMA] = meshgrid(-10:dx:10,1:dsigma:5)}$$

sets up a pair of 17 by 41 two-dimensional arrays X, and SIGMA, that store the σ and x values in a format suitable for the three-dimensional plotting routines. The line

$$\texttt{Z = exp(-(X-mu).\^2./(2*SIGMA.2))./sqrt(2*pi*SIGMA.\^2);}$$

then computes values of the density function. Note that the powering operator, `^`, and the division operator, `/`, are preceded by full stops. This notation allows MATLAB to work directly on arrays by interpreting the commands in a componentwise sense. A simple illustration of this effect is

```
>> [1,2,3].*[5,6,7]
>> ans = 5 12 21
```

The `waterfall` function is then used to give a three-dimensional plot of Z by taking slices along the x-direction. The resulting picture is shown in Figure 3.5.

PROGRAMMING EXERCISES

P3.1. Experiment with `ch03` by varying `dx` and `dsigma`, and replacing `waterfall` by `mesh`, `surf` and `surfc`.

P3.2. Write an analogue of `ch03` for the exponential density function defined in Exercise 3.4.

Quotes

Our intuition is not a viable substitute for the more formal theory of probability.

MARK DENNEY AND STEVEN GAINES (Denney and Gaines, 2000)

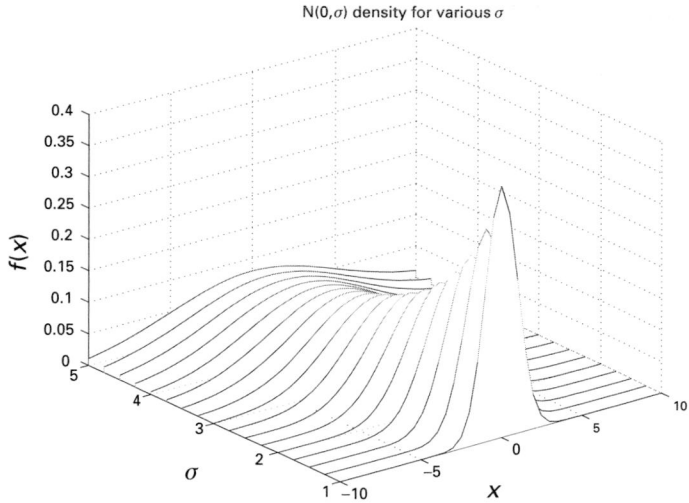

Fig. 3.5. Graphics produced by ch03.

Statistics: the mathematical theory of ignorance.

> MORRIS KLINE, source www.mathacademy.com/pr/quotes/

Stock prices have reached what looks like a permanently high plateau.
(In a speech made nine days before
the 1929 stock market crash.)

> IRVING FISHER, economist, source
> www.quotesforall.com/f/fisherirving.htm

Norman has stumbled into the lair of a chartist,
an occult tape reader who thinks he can predict market moves by eyeballing
the shape that stock prices take when plotted on a piece of graph paper.
Chartists are to finance what astrology is to space science.
It is a mystical practice akin to reading the entrails of animals.
But its newspaper of record is *The Wall Street Journal*,
and almost every major financial institution in the United States
keeps at least one or two chartists working behind closed doors.

> THOMAS A. BASS (Bass, 1999)

4

Computer simulation

4.1 Motivation

The models that we develop for option valuation will involve randomness. One of the main thrusts of this book is the use of *computer simulation* to experiment with and visualize our ideas, and also to estimate quantities that cannot be determined analytically. This chapter introduces the tools that we will apply.

4.2 Pseudo-random numbers

Computers are deterministic – they do exactly what they are told and hence are completely predictable. This is generally a good thing, but it is at odds with the idea of generating random numbers. In practice, however, it is usually sufficient to work with *pseudo-random* numbers. These are collections of numbers that are produced by a deterministic algorithm and yet seem to be random in the sense that, en masse, they have appropriate statistical properties. Our approach here is to assume that we have access to black-box programs that generate large sequences of pseudo-random numbers. Hence, we completely ignore the fascinating issue of designing algorithms for generating pseudo-random numbers. Our justification for this omission is that random number generation is a highly advanced, active, research topic and it is unreasonable to expect non-experts to understand and implement programs that compete with the state-of-the-art. Off-the-shelf is better than roll-your-own in this context, and by making use of existing technology we can more quickly progress to the topics that are central to this book.

Table 4.1. *Ten pseudo-random*
numbers from a $U(0, 1)$ *and*
an $N(0, 1)$ *generator*

$U(0, 1)$	$N(0, 1)$
0.3929	0.9085
0.6398	−2.2207
0.7245	−0.2391
0.6953	0.0687
0.9058	−2.0202
0.9429	−0.3641
0.6350	−0.0813
0.1500	−1.9797
0.4741	0.7882
0.9663	0.7366

Table 4.1 shows two sets of ten numbers. These were produced from high-quality pseudo-random number generators designed to produce $U(0, 1)$ and $N(0, 1)$ samples.[1] We see that the putative $U(0, 1)$ samples appear to be liberally spread across the interval $(0, 1)$ and the putative $N(0, 1)$ samples seem to be clustered around zero, but, of course, this tells us very little.

4.3 Statistical tests

We may test a pseudo-random number generator by taking M samples $\{\xi_i\}_{i=1}^M$ and computing the *sample mean*

$$\mu_M := \frac{1}{M} \sum_{i=1}^M \xi_i, \tag{4.1}$$

and the *sample variance*

$$\sigma_M^2 := \frac{1}{M-1} \sum_{i=1}^M (\xi_i - \mu_M)^2. \tag{4.2}$$

The sample mean (4.1) is simply the arithmetic average of the sample values. The sample variance is a similar arithmetic average corresponding to the expected value in (3.10) that defines the variance. (You might regard it as more natural to take the sample variance as $(1/M) \sum_{i=1}^M (\xi_i - \mu_M)^2$; however, it can be argued that scaling

---

[1] All computational experiments in this book were produced in MATLAB, using the built-in functions `rand` and `randn` to generate $U(0, 1)$ and $N(0, 1)$ samples, respectively. To make the experiments reproducible, we set the random number generator seed to 100; that is, we used `rand('state',100)` and `randn('state',100)`.

Table 4.2. *Sample mean (4.1) and sample variance*
(4.2) using M samples from a U(0, 1) *and an*
N(0, 1) *pseudo-random number generator*

M	U(0, 1)		N(0, 1)	
	μ_M	σ_M^2	μ_M	σ_M^2
10^2	0.5229	0.0924	0.0758	1.0996
10^3	0.4884	0.0845	0.0192	0.9558
10^4	0.5009	0.0833	−0.0115	0.9859
10^5	0.5010	0.0840	0.0005	1.0030

by $M - 1$ instead of M is better. This issue is addressed in Chapter 15.) Results for $M = 10^2, 10^3, 10^4$ and 10^5 appear in Table 4.2. We see that as M increases, the U(0, 1) sample means and variances approach the true values $\frac{1}{2}$ and $\frac{1}{12} \approx 0.0833$ (recall Exercise 3.5) and the N(0, 1) sample means and variances approach the true values 0 and 1.

A more enlightening approach to testing a random number generator is to divide the x-axis into subintervals, or *bins*, of length Δx and count how many samples lie in each subinterval. We take M samples and let N_i denote the number of samples in the bin $[i\Delta x, (i + 1)\Delta x]$. If we approximate the probability of X taking a value in the subinterval $[i\Delta x, (i + 1)\Delta x]$ by the relative frequency with which this happened among the samples, then we have

$$\mathbb{P}(i\Delta x \leq X \leq (i + 1)\Delta x) \approx \frac{N_i}{M}. \tag{4.3}$$

On the other hand, we know from (3.3) that, for a random variable X with density $f(x)$,

$$\mathbb{P}(i\Delta x \leq X \leq (i + 1)\Delta x) = \int_{i\Delta x}^{(i+1)\Delta x} f(x)dx. \tag{4.4}$$

Letting x_i denote the midpoint of the subinterval $[i\Delta x, (i + 1)\Delta x]$ we may use the Riemann sum approximation

$$\int_{i\Delta x}^{(i+1)\Delta x} f(x)dx \approx \Delta x f(x_i). \tag{4.5}$$

(Here, we have approximated the area under a curve by the area of a suitable rectangle – draw a picture to see this.) Using (4.3)–(4.5), we see that plotting $N_i/(M\Delta x)$ against x_i should give an approximation to the density function values

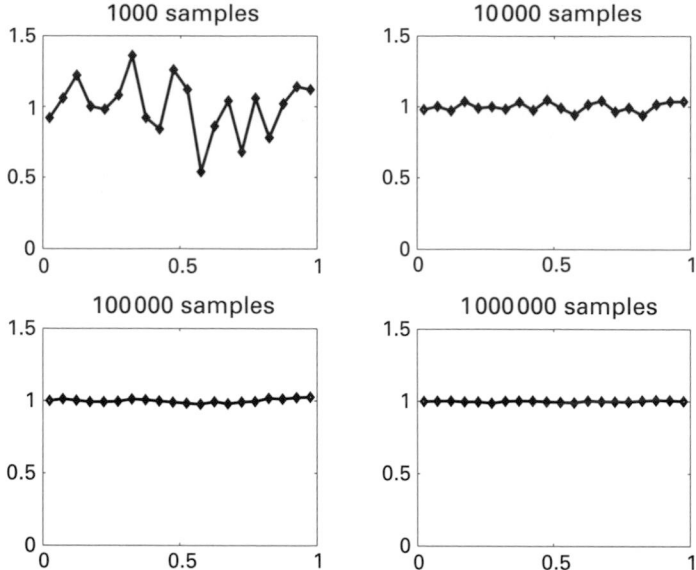

Fig. 4.1.   Kernel density estimate for a $\mathsf{U}(0, 1)$ generator, with increasing number of samples. Vertical axis is $N_i/(M\,\Delta x)$, for $\Delta x = 0.05$.

$f(x_i)$. This technique, and more sophisticated extensions, fit into the area of *kernel density estimation*.

Computational example We compute a simple kernel density estimate for a $\mathsf{U}(0, 1)$ generator, using intervals of width $\Delta x = 0.05$. Since $f(x)$ is nonzero only for $0 \le x \le 1$, we take $i = 0, 1, 2, \ldots, 19$. In Figure 4.1 we plot $N_i/(M\,\Delta x)$ against x_i for the number of samples $M = 10^3, 10^4, 10^5, 10^6$. These points are plotted as diamonds joined by straight lines for clarity. We see that as M increases the plot gets closer to that of a $\mathsf{U}(0, 1)$ density. $\diamond$

Computational example In Figure 4.2 we perform a similar experiment with an $\mathsf{N}(0, 1)$ generator. Here, we took intervals in the region $-4 \le x \le 4$ and used bins of width $\Delta x = 0.05$. (Samples that were smaller than -4 were added to the first bin and samples that were larger than 4 were added to the last bin.) We used $M = 10^3, 10^4, 10^5, 10^6$. The correct $\mathsf{N}(0, 1)$ density curve is superimposed in white. We see that the density estimate improves as M increases. $\diamond$

We now look at another technique for examining statistical aspects of data. For a given density function $f(x)$ and a given $0 < p < 1$ define the *pth quantile* of f as $z(p)$, where

$$\int_{-\infty}^{z(p)} f(x)\,dx = p. \tag{4.6}$$

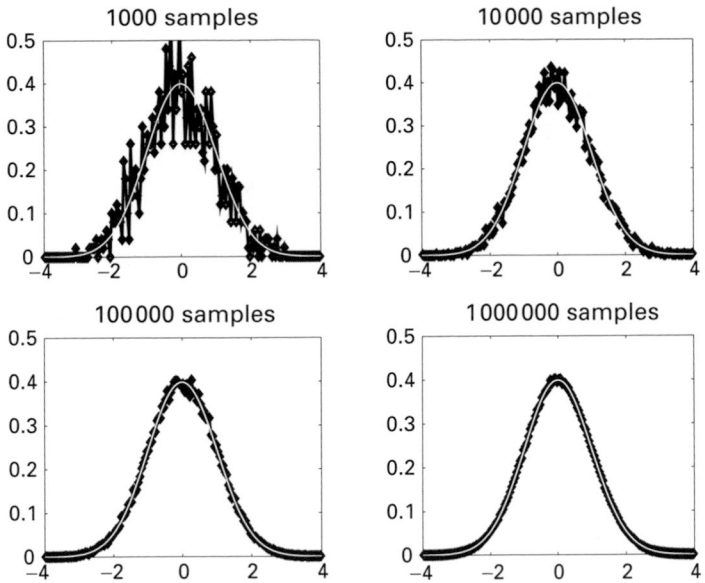

Fig. 4.2. Kernel density estimate for an $N(0, 1)$ generator, with increasing number of samples. Vertical axis is $N_i/(M \Delta x)$, for $\Delta x = 0.05$.

Given a set of data points $\xi_1, \xi_2, \ldots, \xi_M$, a *quantile–quantile plot* is produced by

(a) placing the data points in increasing order: $\widehat{\xi}_1, \widehat{\xi}_2, \ldots, \widehat{\xi}_M$,
(b) plotting $\widehat{\xi}_k$ against $z(k/(M+1))$.

The idea of choosing quantiles for equally spaced $p = k/(M+1)$ is that it 'evens out' the probability. Figure 4.3 illustrates the $M = 9$ case when $f(x)$ is the $N(0, 1)$ density. The upper picture emphasizes that the $z(k/(M+1))$ break the x-axis into regions that give equal area under the density curve – that is, there is an equal chance of the random variable taking a value in each region. The lower picture in Figure 4.3 plots the function $f(x)$ and shows that $z(k/(M+1))$ are the points on the x-axis that correspond to equal increments along the y-axis. The idea is that, for large M, if the quantile–quantile plot produces points that lie approximately on a straight line of unit slope, then we may conclude that the data points 'look as though' they were drawn from a distribution corresponding to $f(x)$. To justify this, if we divide the x-axis into M bins where x is in the kth bin if it is closest to $z(k/(M+1))$, then, having evened out the probability, we would expect roughly one ξ_i value in each bin. So the smallest data point, $\widehat{\xi}_1$, should be close to $z(1/(M+1))$, the second smallest, $\widehat{\xi}_2$, should be close to $z(2/(M+1))$, and so on.

Computational example Figure 4.4 tests the quantile–quantile idea. Here we took $M = 100$ samples from $N(0, 1)$ and $U(0, 1)$ random number generators.

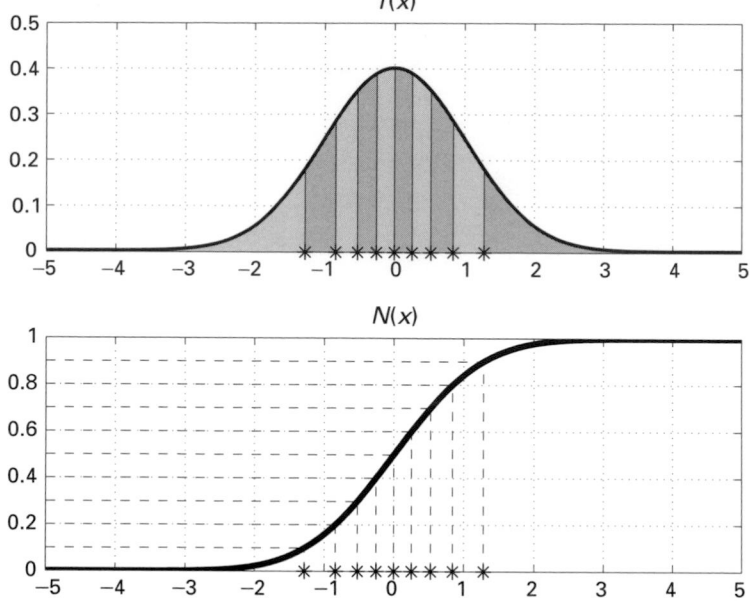

Fig. 4.3.   Asterisks on the x-axis mark the quantiles $z(k/(M+1))$ in (4.6) for an N(0, 1) distribution using $M = 9$. Upper picture: the quantiles break the x-axis into regions where $f(x)$ has equal area. Lower picture: equivalently, the quantiles break the x-axis into regions where $N(x)$ has equal increments.

Each data set was plotted against the N(0, 1) and U(0, 1) quantiles. A reference line of unit slope is added to each plot. As expected, the data set matches well with the 'correct' quantiles and very poorly with the 'incorrect' quantiles.   ◇

Computational example In Figures 4.5 and 4.6 we use the techniques introduced above to show the remarkable power of the Central Limit Theorem. Here, we generated sets of U(0, 1) samples $\{\xi_i\}_{i=1}^{n}$, with $n = 10^3$. These were combined to give samples of the form

$$\frac{\sum_{i=1}^{n} \xi_i - n\mu}{\sigma \sqrt{n}}, \tag{4.7}$$

where $\mu = \frac{1}{2}$ and $\sigma^2 = \frac{1}{12}$. We repeated this $M = 10^4$ times. These M data points were then used to obtain a kernel density estimate. In Figure 4.5 we used bins of width $\Delta x = 0.5$ over $[-4, 4]$ and plotted $N_i/(M\Delta x)$ against x_i, as described for Figure 4.1. Here we have used a *histogram*, or *bar graph*, so each rectangle is centred at an x_i and has height $N_i/(M\Delta x)$. The N(0, 1) density curve is superimposed as a dashed line. Figure 4.6 gives the corresponding quantile–quantile plot. The figures confirm that even though each ξ_i is nothing like normal, the scaled sum $(\sum_{i=1}^{n} \xi_i - n\mu)/(\sigma \sqrt{n})$ is very close to N(0, 1).   ◇

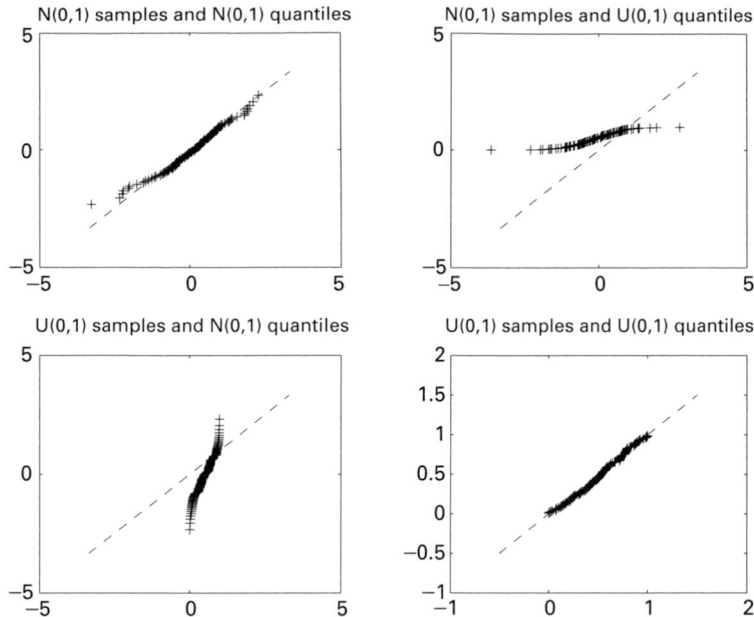

Fig. 4.4. Quantile–quantile plots using $M = 100$ samples. Ordered samples $\widehat{\xi}_1$, $\widehat{\xi}_2, \ldots, \widehat{\xi}_M$ on the x-axis against quantiles $z(k/(M + 1))$ on the y-axis. Pictures show the four possible combinations arising from $N(0, 1)$ or $U(0, 1)$ random number samples against $N(0, 1)$ or $U(0, 1)$ quantiles.

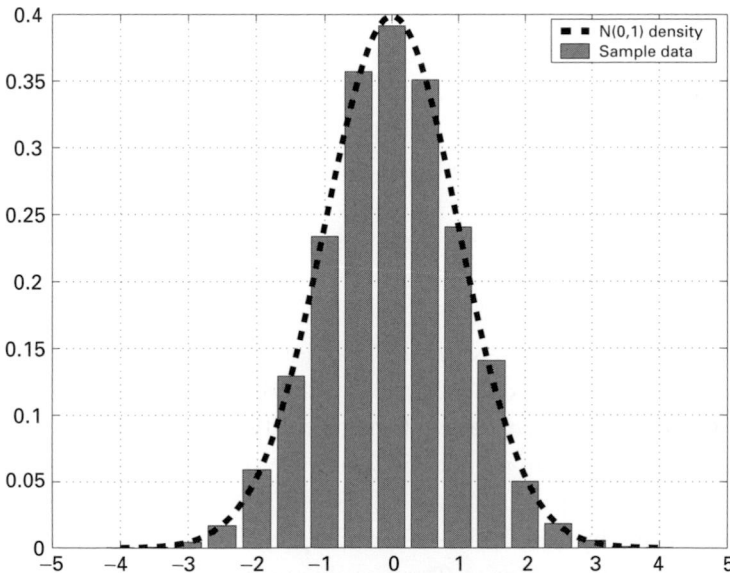

Fig. 4.5. Kernel density estimate for samples of the form (4.7), with $N(0, 1)$ density superimposed.

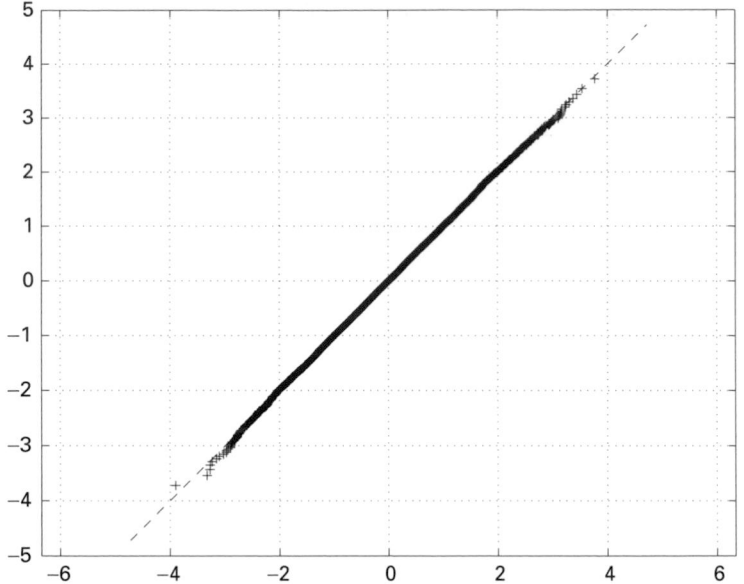

Fig. 4.6.   Quantile–quantile plot for samples of the form (4.7) against $N(0, 1)$ quantiles.

4.4 Notes and references

Much more about the theory and practice of designing and implementing computer simulation experiments can be found in (Morgan, 2000) and (Ripley, 1987). In particular, those references mention rules of thumb for choosing the bin width as a function of the sample size in kernel density estimation.

The pLab website, which lives at http://random.mat.sbg.ac.at/, gives information on random number generation, and has links to free software in a variety of computing languages.

A very readable essay on pseudo-random number generation can be found in (Nahin, 2000). That book also contains some wonderful probability-based problems, with accompanying MATLAB programs.

Cleve's Corner articles 'Normal behavior', spring 2001, and 'Random Thoughts', fall 1995, which are downloadable from www.mathworks.com/company/newsletter/clevescorner/cleve_toc.shtml are informative musings on MATLAB's pseudo-random number generators.

As an alternative to 'pseudo-', it is possible to buy 'true' random numbers that are generated from physical devices. For example, one approach is to record decay times from a radioactive material. The readable article '*Hardware random number generators*', by Robert Davies, can be found at www.robertnz.net/hwrng.htm.

4.1. ⋆⋆ Some scientific computing packages offer a black-box routine to evaluate the *error function*, erf, defined by

$$\text{erf}(x) := \frac{2}{\sqrt{\pi}} \int_0^x e^{-t^2} \, dt. \tag{4.8}$$

Show that the $\mathsf{N}(0, 1)$ distribution function $N(x)$ in (3.18) can be evaluated as

$$N(x) = \frac{1 + \text{erf}\left(x/\sqrt{2}\right)}{2}. \tag{4.9}$$

4.2. ⋆⋆ Show that samples from the exponential distribution with parameter λ, as described in Exercise 3.4, may be generated as $-(\log(\xi_i))/\lambda$, where the $\{\xi_i\}$ are $\mathsf{U}(0, 1)$ samples.

4.3. ⋆⋆ Show that the quantile $z(p)$ in (4.6) for the $\mathsf{N}(0, 1)$ distribution function $N(x)$ can be written as

$$z(p) = \sqrt{2}\,\text{erfinv}(2p - 1).$$

Here, erfinv is the inverse error function; so $\text{erfinv}(x) = y$ means $\text{erf}(y) = x$, where erf is defined in (4.8).

4.4. ⋆⋆ In the case where $f(x)$ is the density for the exponential distribution with parameter $\lambda = 1$, as described in Exercise 3.4, show that the quantile $z(p)$ in (4.6) satisfies $z(p) = -\log(1 - p)$.

4.5 Program of Chapter 4 and walkthrough

In ch04, listed in Figure 4.7, we repeat the type of computation that produced Figure 4.5. Here, we use samples ξ_i in (4.7) that are the exponential of the square root of $\mathsf{U}(0, 1)$ samples. It follows from Exercise 21.2 that we should take $\mu = 2$ and $\sigma = \sqrt{(e^2 - 7)/2}$.

The line `colormap([0.5 0.5 0.5])` sets the greyscale for the histogram; [0 0 0] is black and [1 1 1] is white. We then use `rand('state',100)` to set the seed for the uniform pseudo-random number generator, as described in the footnote of Section 4.2. After specifying n, M, mu and `sigma`, and initializing S to an array of zeros, we perform the main task in a single `for` loop. The command `rand(n,1)` creates an array of n values from the $\mathsf{U}(0, 1)$ pseudo-random number generator. We then apply `sqrt` to take the square root of each entry, `exp` to exponentiate and `sum` to add up the result. In other words

$$\text{sum(exp(sqrt(rand(n,1))))}$$

corresponds to a sample of

$$\sum_{i=1}^{n} e^{\sqrt{\xi_i}},$$

```
%CH04    Program for Chapter 4
%
% Histogram illustration of Central Limit Theorem

clf
colormap([0.5 0.5 0.5])
rand('state',100)

n =  5e+2;
M = 1e+4;
mu = 2;
sigma = sqrt(0.5*(exp(2)-7));

S = zeros(M,1);
for k = 1:M
   S(k) = (sum(exp(sqrt(rand(n,1)))) - n*mu)/(sigma*sqrt(n));
end

%%%%%%%%%%%%%%%%% Histogram %%%%%%%%%%%%%%%%%%%%%%%%%
dx = 0.5;
centers = [-4:dx:4];
N = hist(S,centers);
bar(centers,N/(M*dx))

hold on
x = linspace(-4,4,100);
y = exp(-0.5*x.^2)/sqrt(2*pi);
plot(x,y,'r-','Linewidth',4)
legend('N(0,1) density','Sample data')
grid on
```

Fig. 4.7.   Program of Chapter 4: ch04.m.

so overall, S(k) stores a sample of

$$\frac{\sum_{i=1}^{n} e^{\sqrt{\xi_i}} - n\mu}{\sigma \sqrt{n}}.$$

The line N = hist(S,centers); creates a one-dimensional array N, whose ith entry records the number of values in S lying in the ith bin. Here, a point is mapped to the ith bin if its closest value in centers is centers(i). The command bar(centers,N/(M*dx)) then draws a bar graph, or histogram, using this information. (We scale by M*dx so that the area of the histograph adds up to one.)

Because we issued the command hold on, the second plot, a dashed line for the exact density curve, adds to, rather than replaces, the first.

PROGRAMMING EXERCISES

P4.1. Adapt ch04.m to the case where ξ_i in (4.7) are from the exponential distribution with parameter $\lambda = 1$. [Hint: make use of Exercise 3.4 and Exercise 4.2.]

P4.2. Adapt ch04.m so that it produces a quantile–quantile plot, as in Figure 4.6. (Note that the program of Chapter 5 shows how such a plot may be generated.)

Quotes

In 1955, before computers were so common,
the RAND Corporation published a book entitled *A Million Random Digits*.
It was used in selecting random trials for experimental designs and simulations
(and perhaps as bedtime reading for insomniacs?).
It was soon realized, however, that if everyone always started on page one,
then all trials and simulations by all the book's users would depend upon the quirks of
the same random sequence.
This generated much debate
on how to select a random starting point in the table of random numbers.

MICHAEL T. HEATH (Heath, 2002)

The first thing needed for a stochastic simulation is a source of randomness.
This is often taken for granted but is of fundamental importance.
Regrettably many of the so-called random functions supplied with the most
widespread computers
are far from random,
and many simulation studies have been invalidated as a consequence.

BRIAN D. RIPLEY (Ripley, 1997)

Here is an interesting number:
0.950 129 285 147 18.
This is the first number produced by the MATLAB random number generator with its
default settings.
Start up a fresh MATLAB, set format long, type rand,
and it's the number you get.
If all MATLAB users, all around the world, on all different computers,
keep getting this same number, is it really 'random'?
No, it isn't.
Computers are (in principle) deterministic machines
and should not exhibit random behavior.
If your computer doesn't access some external device,
like a gamma ray counter or a clock,
then it must really be computing *pseudorandom* numbers.

CLEVE B. MOLER AND KATHRYN A. MOLER, in *Numerical Computing with MATLAB*,
see www.mathworks.com/moler/

5

Asset price movement

5.1 Motivation

In order to value an option, we must develop a mathematical description of how the underlying asset behaves. This chapter gives examples of real stock market data and performs some basic statistical tests. The tests pave the way for the mathematical description that we introduce in the next chapter, but are definitely not intended to form an exhaustive justification of the model. We begin with an outline of a key hypothesis, and finish by listing some of the assumptions that will go into our analysis.

5.2 Efficient market hypothesis

The price of an asset is, of course, a measure of investors' confidence, and, as such, is strongly dependent upon news, rumours, speculation, and so on. Although an oversimplification, it is reasonable to assume that the market responds instantaneously to external influences, and hence:

the current asset price reflects all past information.

This simple conclusion is known as the (weak form of the) *efficient market hypothesis*. Under this hypothesis, if we want to predict the asset price at some future time, knowing the complete history of the asset price gives no advantage over just knowing its current price – there is no edge to be gained from 'reading the charts.'

45

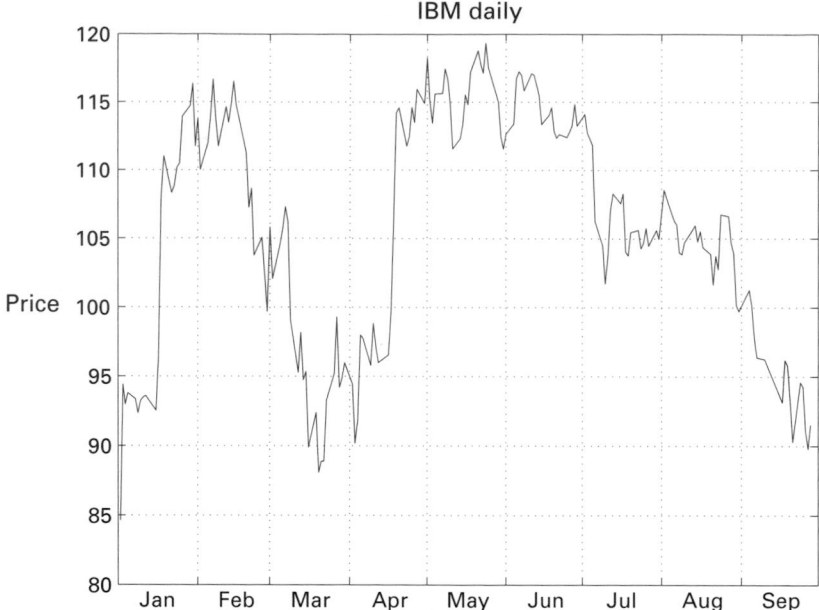

Fig. 5.1. Daily IBM share price from January to September 2001.

From a modelling point of view, if we take on board the efficient market hypothesis, then an equation to describe the evolution of the asset from time t to $t + \Delta t$ need involve the asset price only at time t and **not** at any earlier times.

5.3 Asset price data

In Figure 5.1 we plot the daily IBM share prices from January to the end of September 2001. These are the close-of-trading prices; that is, the price at the last transaction made in each trading day. In the traditional manner, we have 'joined the dots' so that successive data points are linked by straight lines. Figure 5.2 gives the corresponding weekly IBM share prices from January 1998 to December 2001. There are 184 data points in Figure 5.1 and 209 in Figure 5.2. Although covering different timescales, both pictures display the same qualitative 'jaggedness'. This type of up/down uncertainty is familiar to anybody who has seen stock market data displayed in graphical form.

To examine this data, it is reasonable to treat it on the same level as the output from a pseudo-random number generator and test whether it has any statistical properties. In Figure 5.3 we give the results of such a test. The upper pictures involve the *daily returns*,

$$r_i^{\text{daily}} := \frac{S(t_{i+1}) - S(t_i)}{S(t_i)},$$

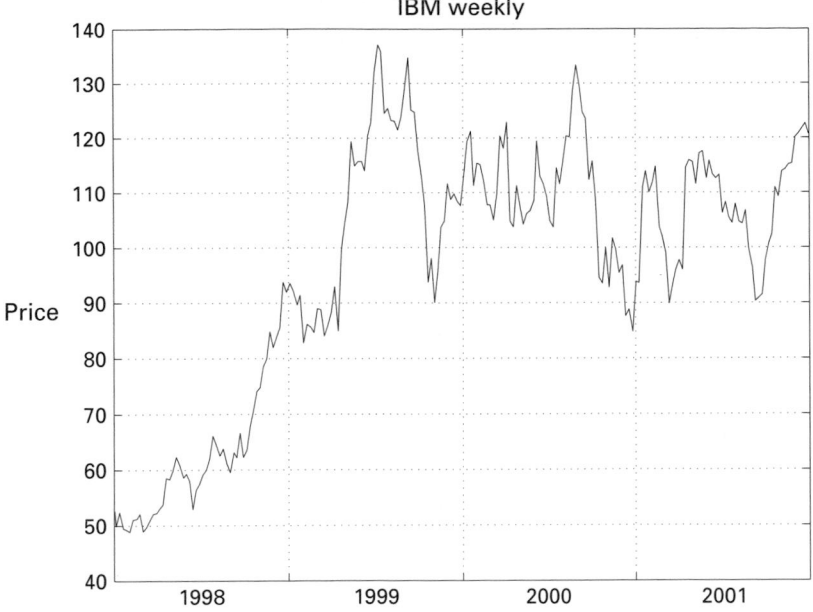

Fig. 5.2. Weekly IBM share price from January 1998 to December 2001.

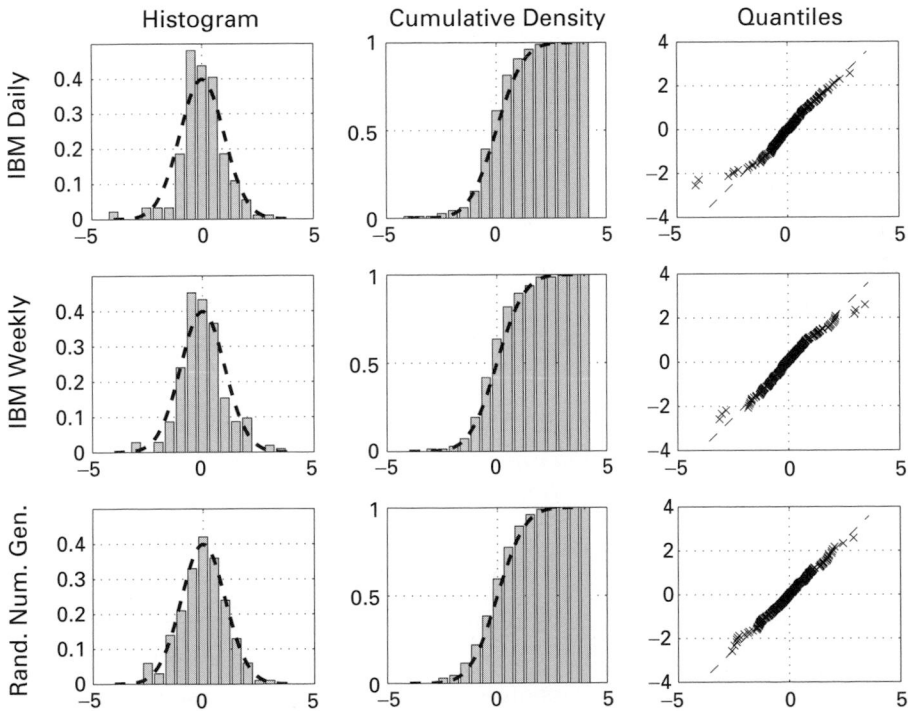

Fig. 5.3. Statistical tests of IBM share price data. Upper: daily. Middle: weekly. Lower: $N(0, 1)$ samples for comparison.

where $S(t_i)$ and $S(t_{i+1})$ are the asset prices on successive days, as used in Figure 5.1. These daily returns were normalized to

$$\widehat{r}_i^{\,\text{daily}} := \frac{r_i^{\,\text{daily}} - \mu}{\sigma},$$

where μ and σ^2 are the computed sample mean and sample variance, defined in (4.1) and (4.2), respectively. If the daily return data looks like i.i.d. samples from a normal distribution, then $\widehat{r}_i^{\,\text{daily}}$ will look like i.i.d. $\mathsf{N}(0, 1)$ samples. The upper left picture in Figure 5.3 gives a kernel density estimate for the $\widehat{r}_i^{\,\text{daily}}$ data in the form of a histogram, with the $\mathsf{N}(0, 1)$ density curve (3.15) superimposed as a dashed line. To estimate the corresponding distribution function, we may use a cumulative sum histogram, where in each bin we record the proportion of samples that fall in that bin, or in a bin to the left. This produces the histogram in the middle picture. The $\mathsf{N}(0, 1)$ distribution function (3.18) is superimposed as a dashed line. Finally, in the upper right picture we give a quantile–quantile plot, as described in Chapter 4, using $\mathsf{N}(0, 1)$ quantiles. The three middle pictures in Figure 5.3 present the same results for the normalized weekly returns, using the data from Figure 5.2. As a basis for comparison, the lower pictures give the output that arises when 200 points from an $\mathsf{N}(0, 1)$ pseudo-random number generator are subjected to the same scrutiny.

Overall, Figure 5.3 suggests that the daily and weekly asset returns behave in a similar manner to normally distributed i.i.d. samples. The quantile–quantile plots, which are the most revealing, possibly indicate that the match is least accurate at the extremes of the range – this *fat tail* behaviour will be mentioned again in Section 7.4.

As a final point, we remark that since the daily and weekly returns are quite small, the approximation $\log(1 + x) \approx x$ gives

$$\log\left(\frac{S(t_{i+1})}{S(t_i)}\right) = \log\left(1 + \frac{S(t_{i+1}) - S(t_i)}{S(t_i)}\right) \approx \frac{S(t_{i+1}) - S(t_i)}{S(t_i)} \qquad (5.1)$$

and hence we would see essentially the same pictures as those in Figure 5.3 if we replaced the returns with the *log ratios*, $\log\left(S(t_{i+1})/S(t_i)\right)$.

5.4 Assumptions

In the next chapter we develop a mathematical description of the asset price movement that is intended to capture the broad features that are observed in practice. Before we do that, we take the opportunity to list some of the assumptions that will be made in the subsequent analysis.

- The asset price may take any non-negative value.
- Buying and selling an asset may take place at any time $0 \le t \le T$.
- It is possible to buy and sell any amount of the asset.
- The bid–ask spread is zero – the price for buying equals the price for selling.
- There are no transaction costs.
- There are no dividends or stock splits.
- Short selling is allowed – it is possible to hold a negative amount of the asset.
- There is a single, constant, risk-free interest rate that applies to any amount of money borrowed from or deposited in a bank.

5.5 Notes and references

The efficient market hypothesis is at best an approximation to reality. A classic text that espouses the hypothesis is (Malkiel, 1990). A more recent book that analyses vast amounts of stock market data and casts severe doubt on the efficient market hypothesis is (Lo and MacKinlay, 1999). It is important to keep in mind, however, that it is a big leap to go from

(a) claiming that the current asset price movement is somehow correlated with historical asset price data, to
(b) developing a method that can make these correlations sufficiently explicit to be of use for prediction.

Bass (Bass, 1999) describes what seems to be one of the few successful, systematic attempts in this direction. The topic is mentioned further in Section 7.4.

The data used in Figures 5.1–5.3 was downloaded from the Yahoo! Finance website at http://finance.yahoo.com/ and processed using MATLAB code based on the tools developed by Petter Wiberg at www.maths.warwick.ac.uk/wiberg/MathFinance/.

It is worth emphasizing that the tests in Section 5.3 were designed solely for the purpose of illustration. There are many practical issues to address before a serious statistical analysis of stock market data can be performed. Most notably:

- There may be *missing data* if no trading took place between times t_i and t_{i+1}.
- For many data sets, each price may correspond to either a buy or a sell – there is an in-built noise level at the order of the bid–ask spread.
- The data may require adjustments to account for dividends and stock splits.
- When determining the time interval, $t_{i+1} - t_i$, between price data, a decision must be made about whether to keep the clock running when the stock market has closed. Does Friday night to Monday morning count as $2\frac{1}{2}$ days, or zero days?
- For an asset that is not heavily traded, the time of the last trade may vary considerably from day to day. Consequently, daily closing prices, which pertain to the final trade for each day, may not relate to equally spaced samples in time.

The book (Lo and MacKinlay, 1999) is a good source of practical information for stock market data analysis.

Many exchanges have informative websites, including **the American Stock Exchange:** www.amex.com/, **the Chicago Board Options Exchange:** www. cboe.com/Home/, **the London Stock Exchange:** www.londonstockexchange. com/, **the New York Stock Exchange:** www.nyse.com/.

EXERCISES

5.1. ⋆⋆ Consider the following quote from Eugene Fama, who was Myron Scholes' thesis adviser, which can be found in (Lowenstein, 2001, page 71).

> If the population of price changes is strictly normal, on the average for any stock
> … an observation more than five standard deviations from the mean should be observed about once every 7000 years. In fact such observations seem to occur about once every three to four years.

Given that for $X \sim N(\mu, \sigma^2)$, $\mathbb{P}(|X - \mu| > 5\sigma) = 5.733 \times 10^{-7}$, deduce how many observations per year Fama is implicitly assuming to be made.

5.2. Complete the following stock market report in an apt and amusing manner.

- Knives fell sharply.
- Guacamole dipped.
- Toilet tissue bottomed out ....

5.6 Program of Chapter 5 and walkthrough

The program ch05 shows one way to compute a quantile–quantile plot, as seen in Figures 4.4, 4.6 and 5.3. It is listed in Figure 5.4. We use MATLAB's $N(0, 1)$ pseudo-random number generator, randn. The line samples = randn(M,1), assigns M such samples to the array samples. We then use ssort = sort(sample), to create an array ssort containing the elements of samples, rearranged into ascending order. The line pvals = [1:M]/(M+1), then sets up equally spaced points $1/(M + 1), 2/(M + 1), 3/(M + 1), \ldots, M/(M + 1)$ and zvals = sqrt(2)*erfinv(2*pvals-1); computes the required quantiles, as described in Exercise 4.3. We then plot the ordered samples against the quantiles and superimpose a reference line of slope one.

PROGRAMMING EXERCISES

P5.1. Use the cumulative sum function cumsum and the bar graph function bar to produce a cumulative density plot from ch05.m, as in the lower middle picture of Figure 5.3.

P5.2. Use the code at www.maths.warwick.ac.uk/wiberg/MathFinance/ to manipulate and display real stock market data.

```
%CH05  Program for Chapter 5
%
% Illustrates quantile plot

clf
randn('state',100)

M = 200;

samples = randn(M,1);
ssort = sort(samples);

pvals = [1:M]/(M+1);
zvals =  sqrt(2)*erfinv(2*pvals-1);

plot(ssort,zvals,'rx')
hold on
xlim  = max(abs(zvals))+1;
plot([-xlim, xlim],[-xlim,xlim],'g–') % Reference of slope 1
title('N(0,1) quantile-quantile plot')
grid on
```

Fig. 5.4.   Program of Chapter 5: ch05.m.

Quotes

A battle rages between those who say the financial markets
are theoretically impossible to beat and those who say,
'Hey, look at me, I'm a billionaire.'
On one side are the Nobel laureates,
ensconced in the University of Chicago Business School,
who are renowned for developing equations describing 'efficient',
that is, unbeatable, markets.
On the other side are the speculators who beat them year in, year out
with techniques 'proven' not to work.

THOMAS A. BASS (Bass, 1999)

Who'd have imagined that our largest single equity underwriting
would coincide with the largest drop in history in the stock market?
Then, who'd have imagined that our first big junk bond deal
would coincide with the crash of the junk bond market?
It was striking how little control we had of events,
particularly in view of how assiduously
we cultivated the appearance of being in charge
by smoking big cigars and saying **** all the time.

MICHAEL LEWIS (Lewis, 1989)

An incident of 'fat finger syndrome' – inadvertently pressing the wrong button
on a computer keyboard – landed an American investment bank

with multimillion pound losses yesterday
and is expected to cost the young city trader involved his job.
. . . The deal amounted to £300m rather than £3m
and flashed across stock market screens just as the stock market was about to close,
causing a precipitous fall in the Footsie, the barometer of British corporate health.

Slip of the finger that cost city dearly, the *Guardian*, 16 May 2001

The traditional view in economics
is that financial agents are completely rational with perfect foresight.
Markets are always in equilibrium,
which in economics means that trading always occurs
at a price that conforms to everyone's expectations of the future.
Markets are efficient, which means that there are no patterns in prices
that can be forecast based on a given information set.
The only possible changes in price are random,
driven by unforecastable external information.
Profits occur only by chance.
In recent years this view is eroding.

J. DOYNE FARMER (Farmer, 1999)

6

Asset price model: Part I

6.1 Motivation

Our aim in this chapter is to motivate and derive the classic model for asset price behaviour. We do this in a heuristic manner, making clear the assumptions that are being made and keeping in mind that the model will be used as the basis for an option valuation theory.

Given the asset price S_0 at time $t = 0$, our objective is to come up with a process that describes the asset price $S(t)$ for all times $0 \leq t \leq T$. Due to the unpredictable nature of assset price movements, $S(t)$ will be a random variable for each t. Although asset prices are typically rounded to one or two decimal places, we assume here that an asset may have any price ≥ 0.

Our approach is to set up an expression for the relative change over an interval of time δt and then let $\delta t \to 0$ in order to get an expression that is valid for continuous t.

6.2 Discrete asset model

As a starting point for our model we note from Exercise 2.2 that the change in the value of a risk-free investment over a small time interval δt can be modelled as

$$D(t + \delta t) = D(t) + r\delta t D(t), \tag{6.1}$$

where r is the interest rate. In order to account for the typical, unpredictable changes in asset price, we will add a random element to this equation. We saw in Chapter 5 that the efficient market hypothesis says that the current asset price reflects all the information known to investors, and hence any change in the price is due to new information. We may build this into our model by adding a random 'fluctuation' increment to the interest rate equation and making these increments **independent** for different subintervals. To make this precise, let $t_i = i\delta t$, so that asset prices are to be determined at discrete points $\{t_i\}$. (We will then let $\delta t \to 0$ to get an asset price model over $0 \le t \le T$.) Our discrete-time model is

$$S(t_{i+1}) = S(t_i) + \mu \delta t S(t_i) + \sigma \sqrt{\delta t}\, Y_i S(t_i), \qquad (6.2)$$

where

- μ is a constant parameter. (Typically $\mu > 0$, so that $\mu \delta t S(t_i)$ represents a general upward drift of the asset price. The parameter μ plays the same role as the interest rate r in (6.1).)
- $\sigma \ge 0$ is a constant parameter that determines the strength of the random fluctuations.
- $Y_0, Y_1, Y_2, \ldots$ are i.i.d. $\mathsf{N}(0, 1)$.

  It is worth emphasizing a few points.

 (i) Since a $\mathsf{N}(0, 1)$ random variable is symmetric about the origin, the fluctuation factor $\sigma \sqrt{\delta t} Y_i$ is equally likely to be positive or negative, and the probability that it lies in an interval $[a, b]$ is the same as the probability that it lies in the interval $[-b, -a]$.
 (ii) The presence of the factor $\sqrt{\delta t}$ (rather than some other power of δt) turns out to be necessary in order for a sensible continuous-time limit to exist. Exercise 6.1 follows this through.
(iii) The choice of a normal distribution for Y_i is not arbitrary – because of the Central Limit Theorem, we would arrive at the same continuous-time model for $S(t)$ if we just assumed that $\{Y_i\}_{i \ge 0}$ were i.i.d. with zero mean and unit variance. Exercise 6.2 asks you to confirm this.

  The parameter μ in (6.2) is usually called the *drift* and σ is called the *volatility*. The model is statistically the same if σ is replaced by $-\sigma$, see Exercise 6.3. Convention dictates that σ is taken to be ≥ 0. Typical values for σ lie between 0.05 and 0.5, that is, 5% and 50% volatility. Because we are measuring time in years, the units of σ^2 are per annum. The drift parameter is typically between 0.01 and 0.1, but, as we will see in Chapter 8, its value turns out to be irrelevant in valuing an option.

  We point out that in the model (6.2), the returns $(S(t_{i+1}) - S(t_i))/S(t_i)$ form a normal i.i.d. sequence, in line with the broad conclusions that we drew in Section 5.3 after examining real data.

6.3 Continuous asset model

Suppose we consider the time interval $[0, t]$ with $t = L\delta t$. We know $S(0) = S_0$ and the discrete model (6.2) gives us expressions for $S(\delta t), S(2\delta t), \ldots, S(L\delta t = t)$. The plan is to let $\delta t \to 0$, and hence let $L \to \infty$, to get a limiting expression for $S(t)$.

The discrete model (6.2) says that over each δt time interval the asset price gets multiplied by a factor $1 + \mu\delta t + \sigma\sqrt{\delta t}Y_i$, and hence

$$S(t) = S_0 \prod_{i=0}^{L-1} \left(1 + \mu\delta t + \sigma\sqrt{\delta t}Y_i\right).$$

Dividing through by S_0 and taking logs gives

$$\log\left(\frac{S(t)}{S_0}\right) = \sum_{i=0}^{L-1} \log(1 + \mu\delta t + \sigma\sqrt{\delta t}Y_i). \tag{6.3}$$

We are interested in the limit $\delta t \to 0$, so we would like to exploit the approximation $\log(1 + \epsilon) \approx \epsilon - \epsilon^2/2 + \cdots$, for small ϵ. There is a technical issue that we will gloss over. The quantity Y_i in (6.2) is a random variable, not just a real number, but it can be shown that what we are about to do is justifiable because $\mathbb{E}(Y_i^2)$ is finite. Continuing in the belief that the log expansion remains valid, we obtain

$$\log\left(\frac{S(t)}{S_0}\right) \approx \sum_{i=0}^{L-1}(\mu\delta t + \sigma\sqrt{\delta t}Y_i - \tfrac{1}{2}\sigma^2\delta t Y_i^2), \tag{6.4}$$

where we have ignored terms that involve the power $\delta t^{3/2}$ or higher. Exercise 6.4 asks you to show that

$$\mathbb{E}\left(\mu\delta t + \sigma\sqrt{\delta t}Y_i - \tfrac{1}{2}\sigma^2\delta t Y_i^2\right) = \mu\delta t - \tfrac{1}{2}\sigma^2\delta t \tag{6.5}$$

and

$$\text{var}\left(\mu\delta t + \sigma\sqrt{\delta t}Y_i - \tfrac{1}{2}\sigma^2\delta t Y_i^2\right) = \sigma^2\delta t + \text{higher powers of } \delta t. \tag{6.6}$$

Now, insight from the Central Limit Theorem suggests that $\log(S(t)/S_0)$ in (6.4) will behave like a normal random variable with mean $L(\mu\delta t - \tfrac{1}{2}\sigma^2\delta t) = (\mu - \tfrac{1}{2}\sigma^2)t$ and variance $L\sigma^2\delta t = \sigma^2 t$, that is, approximately,

$$\log\left(\frac{S(t)}{S_0}\right) \sim \mathsf{N}((\mu - \tfrac{1}{2}\sigma^2)t, \sigma^2 t). \tag{6.7}$$

Based on these arguments, our limiting *continuous-time* expression for the asset price at time t becomes

$$S(t) = S_0 e^{(\mu - \frac{1}{2}\sigma^2)t + \sigma\sqrt{t}Z}, \qquad \text{where } Z \sim \mathsf{N}(0, 1). \qquad (6.8)$$

In this derivation there was nothing special about starting at time zero – we can equally well argue that the asset price evolves from time $t = t_1$ to $t = t_2$, where $t_2 > t_1$, according to

$$\log\left(\frac{S(t_2)}{S(t_1)}\right) \sim \mathsf{N}\big((\mu - \tfrac{1}{2}\sigma^2)(t_2 - t_1), \sigma^2(t_2 - t_1)\big).$$

A key point is that across non-overlapping time intervals, the normal random variables that describe these changes will be **independent**. This follows because the Y_i in (6.2) are i.i.d. Hence, for $t_3 > t_2 > t_1$ we have

$$\log\left(\frac{S(t_3)}{S(t_2)}\right) \sim \mathsf{N}\big((\mu - \tfrac{1}{2}\sigma^2)(t_3 - t_2), \sigma^2(t_3 - t_2)\big),$$

and is independent of $\log\left(\dfrac{S(t_2)}{S(t_1)}\right)$.

So we can describe the evolution of the asset over any sequence of time points $0 = t_0 < t_1 < t_2 < t_3 < \cdots < t_M$ by

$$S(t_{i+1}) = S(t_i)e^{(\mu - \frac{1}{2}\sigma^2)(t_{i+1} - t_i) + \sigma\sqrt{t_{i+1} - t_i}\,Z_i}, \qquad \text{for i.i.d. } Z_i \sim \mathsf{N}(0, 1). \quad (6.9)$$

6.4 Lognormal distribution

A random variable $S(t)$ of the form (6.8) has a so-called *lognormal* distribution; that is, its log is normally distributed. Note from (6.8) that since $S_0 > 0$, $S(t)$ is guaranteed to be positive at any time; we have $\mathbb{P}(S(t) > 0) = 1$, for any $t > 0$. So $S(t)$ takes values in $(0, \infty)$. The corresponding density function for $S(t)$ is

$$f(x) = \frac{\exp\left(\frac{-(\log(x/S_0) - (\mu - \sigma^2/2)t)^2}{2\sigma^2 t}\right)}{x\sigma\sqrt{2\pi t}}, \qquad \text{for } x > 0, \qquad (6.10)$$

with $f(x) = 0$, for $x \leq 0$, see Exercise 6.5.

The expected value, second moment, and variance of $S(t)$ with this model turn out to be

$$\mathbb{E}(S(t)) = S_0 e^{\mu t}, \qquad (6.11)$$

$$\mathbb{E}(S(t)^2) = S_0^2 e^{(2\mu + \sigma^2)t}, \qquad (6.12)$$

$$\mathsf{var}(S(t)) = S_0^2 e^{2\mu t}(e^{\sigma^2 t} - 1), \qquad (6.13)$$

see Exercise 6.6.

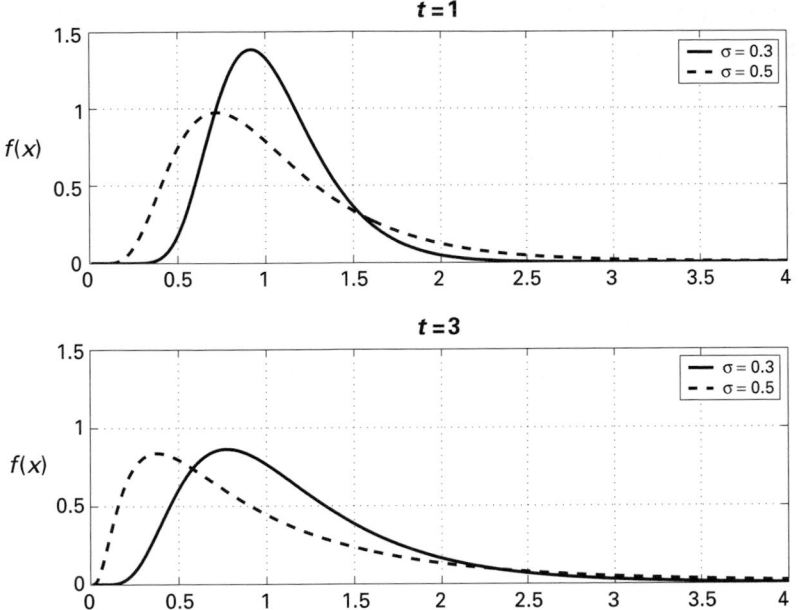

Fig. 6.1. Lognormal density (6.10) for $\mu = 0.05$, $S_0 = 1$, with $\sigma = 0.3$ (solid) and $\sigma = 0.5$ (dashed). Upper picture $t = 1$. Lower picture $t = 3$.

Computational example In Figure 6.1 we set $S_0 = 1$ and $\mu = 0.05$, and plot the lognormal density function (6.10) for $\sigma = 0.3$ and $\sigma = 0.5$. The upper picture is for $t = 1$ and the lower picture for $t = 3$. Note that the density is skewed – it has no vertical axis of symmetry. We know from (6.13) that the variance of $S(t)$ grows with t, and this is clear from the figure – the density function spreads out when t increases. The mean of $S(t)$ also grows with t, from (6.11), although this is less obvious in the figure. $\diamond$

We are deliberately avoiding the direct use of stochastic calculus in this book. However, it is worth mentioning that the process $S(t)$ defined by (6.9) can be regarded as the solution of a *stochastic differential equation* (SDE). In this context, $S(t)$ is often referred to as *geometric Brownian motion*. Section 6.6 gives some routes into this fascinating topic.

6.5 Features of the asset model

We can get some feeling for a continuous random variable by examining its confidence intervals. Suppose that

$$\mathbb{P}\left(a \leq X \leq b\right) = 0.95.$$

Then we say that $[a, b]$ is a *95% confidence interval* for X. In the case where X is normal, there is no simple formula for the inverse of the distribution function $N(x)$ in (3.18), and hence confidence intervals must be computed numerically. It is found that for $X \sim N(0, 1)$,

$$\mathbb{P}\left(|X| \leq 1.96\right) = 0.95, \tag{6.14}$$

see Exercise 6.7, so $[-1.96, 1.96]$ is a 95% confidence interval for X. More generally, for $X \sim N(\mu, \sigma^2)$, we have $(Y - \mu)/\sigma \sim N(0, 1)$, so

$$\mathbb{P}\left(\mu - 1.96\sigma \leq Y \leq \mu + 1.96\sigma\right) = 0.95, \tag{6.15}$$

and hence $[\mu - 1.96\sigma, \mu + 1.96\sigma]$ is a 95% confidence interval. This result is often expressed along the lines of

for i.i.d. normal samples, 95 times out of 100 the sample lies within two standard deviations of the mean.

It follows from (6.7) that

$$[S_0 e^{-1.96\sigma \sqrt{t}+(\mu-\frac{1}{2}\sigma^2)t}, \; S_0 e^{1.96\sigma \sqrt{t}+(\mu-\frac{1}{2}\sigma^2)t}] \tag{6.16}$$

is a 95% confidence interval for the asset price $S(t)$, see Exercise 6.9. If t is small, then

$$e^{-1.96\sigma \sqrt{t}+(\mu-\frac{1}{2}\sigma^2)t} \approx e^{-1.96\sigma \sqrt{t}} \approx 1 - 1.96\sigma \sqrt{t}$$

and

$$e^{1.96\sigma \sqrt{t}+(\mu-\frac{1}{2}\sigma^2)t} \approx e^{1.96\sigma \sqrt{t}} \approx 1 + 1.96\sigma \sqrt{t}.$$

So the confidence interval is approximately

$$[S_0(1 - 1.96\sigma \sqrt{t}), \; S_0(1 + 1.96\sigma \sqrt{t})].$$

The width of this interval is $2S_0 1.96\sigma \sqrt{t}$. If we regard the confidence interval width as a measure of the uncertainty in the future asset price, then this result explains the traders' rule-of-thumb that

over small time periods, uncertainty grows like the square root of time.

Although option valuation is concerned only with the asset price over a fixed time horizon, $[0, T]$, it is interesting to see what the model (6.8) predicts about long term behaviour. Since μ and σ are positive, we see from (6.12) that

$$\lim_{t \to \infty} \mathbb{E}(S(t)^2) = \infty, \qquad \text{as} \quad t \to \infty.$$

In words, we say that the asset tends to infinity in *mean square* as time increases. On the other hand, it can be shown that the $(\mu - \frac{1}{2}\sigma^2)t$ term dominates the $\sigma \sqrt{t} Z$

term in (6.8), so that, with probability 1,

$$\lim_{t \to \infty} S(t) = \begin{cases} \infty, & \text{if} \quad \mu - \frac{1}{2}\sigma^2 > 0, \\ 0, & \text{if} \quad \mu - \frac{1}{2}\sigma^2 < 0. \end{cases} \tag{6.17}$$

So, according to the model, if the volatility is sufficiently large ($\sigma^2 > 2\mu$) then, with probability 1, the asset price will eventually decay to zero.

6.6 Notes and references

The asset price model that we developed is extremely widely used in mathematical finance. The discrete version (6.2) can be regarded as a *numerical approximation* to the SDE formulation. The text (Kloeden and Platen, 1992) is the classic in this area. The expository articles (Higham, 2001; Higham and Kloeden, 2002) give lower level entry points.

The continuous model characterized by (6.8) and (6.9) is the solution to an SDE. Reasonably accessible SDE texts are (Gard, 1988; Mao, 1997; Øksendal, 1998), although all require some background in stochastic processes – the text (Brzeźniak and Zastawniak, 1999) is a good place for beginners to start.

The result (6.17) can be established through the *Strong Law of Large Numbers*, and the $\mu = \frac{1}{2}\sigma^2$ case can be dealt with by the *Law of the Iterated Logarithm*; these laws are discussed, for example, in (Grimmett and Stirzaker, 2001; Kloeden and Platen, 1999).

Although widely used, the lognormal asset price model is, of course, extremely simplistic and open to criticism. Section 7.4 gives pointers to some of the work that has been done on alternative models.

EXERCISES

6.1. ★★ Consider the following variations on the discrete model:

$$S(t_{i+1}) = S(t_i) + \mu \delta t \, S(t_i) + \sigma \delta t \, Y_i \, S(t_i),$$

and

$$S(t_{i+1}) = S(t_i) + \mu \delta t \, S(t_i) + \sigma \delta t^{\frac{1}{4}} Y_i \, S(t_i).$$

By mimicking the heuristic derivation that led to the continuous model (6.8), show that neither of these two variations is satisfactory.

6.2. ★★★ Consider the discrete model (6.2) in the case where $\{Y_i\}$ are general i.i.d. random variables with zero mean and unit variance (i.e., not necessarily normal). Assume also that $\mathbb{E}(Y_i^3)$ and $\mathbb{E}(Y_i^4)$ are finite. Mimic the heuristic derivation that led to (6.8) and show that the same continuous model arises.

6.3.   ⋆⋆ Explain why the model (6.2) is 'statistically the same if σ is replaced by $-\sigma$'.

6.4.   ⋆⋆ Verify (6.5) and (6.6). [Hint: use Exercise 3.7.]

6.5.   ⋆⋆⋆ Show that $S(t)$ in (6.8) has density function (6.10). [Hint: use the characterization $\mathbb{P}(a \leq S(t) \leq b) = \int_a^b f(x)\, dx$ from (3.3).]

6.6.   ⋆⋆⋆ Using (3.8), show that (6.11), (6.12) and (6.13) follow from (6.10).

6.7.   ⋆⋆ Let α be the number such that

$$\mathbb{P}\left(|Z| \leq \alpha\right) = 0.95, \qquad \text{where } Z \sim \mathsf{N}(0, 1).$$

Recalling that $N(\cdot)$ denotes the $\mathsf{N}(0, 1)$ distribution function, show that α satisfies

$$N(-\alpha) = \frac{0.05}{2}.$$

After referring to Exercise 13.3, show that α satisfies $\alpha = \sqrt{2}$ `erfinv` `(0.95)`. Typing this into MATLAB gives $\alpha = 1.9600$ (to 4 decimal places).

6.8.   ⋆ Given that $\mathbb{P}\left(|Z| \leq 2.58\right) = 0.99$, for $Z \sim \mathsf{N}(0, 1)$, show how (6.14) changes when a 99% confidence interval is required.

6.9.   ⋆ Show from (6.7) that (6.16) gives a 95% confidence interval for the asset price $S(t)$.

6.10.   ⋆ Using Exercise 6.8, derive a 99% confidence interval for the asset price $S(t)$. Does the traders' rule-of-thumb still apply?

6.7 Program of Chapter 6 and walkthrough

In ch06, listed in Figure 6.2, we plot the lognormal density function for two different σ values. The resulting picture is similar to those in Figure 6.1. The array y1 stores the value of the density function at equally spaced points in x for the first set of parameter values: t = 1, S = 1, mu = 0.05 and sigma = 0.3. We plot the curve as a red dashed line. The computation is then repeated with sigma = 0.5 and a blue dotted curve is drawn.

PROGRAMMING EXERCISES

P6.1. Adapt ch06 to give a `waterfall` plot illustrating how the lognormal density function varies with σ for fixed $t = 1$.

P6.2. Repeat programming exercise P6.1 for t varying and σ fixed at 1.

```
%CH06    Program for Chapter 6
%
% Plots lognormal density function.

clf

x = linspace(.01,4,500);
t = 1; S = 1; mu = 0.05;

sigma = 0.3;
tempa = ((log(x/S) - (mu-0.5*sigma^2)*t).^2)/(2*t*sigma^2);
tempb = x*sigma*sqrt(2*pi*t);
y1 = exp(-tempa)./tempb;
plot(x,y1,'r-')
ylim([0 1.5])
hold on

sigma = 0.5;
tempa = ((log(x/S) - (mu-0.5*sigma^2)*t).^2)/(2*t*sigma^2);
tempb = x*sigma*sqrt(2*pi*t);
y2 = exp(-tempa)./tempb;
plot(x,y2,'b:')

legend('\sigma = 0.3','\sigma = 0.5',1)
title('Lognormal density, t = 1, S=1, \mu = 0.05')
xlabel('x'), ylabel('f(x)')
```

Fig. 6.2.   Program of Chapter 6: ch06.m.

Quotes

The authors emphasize that,
as even the most cursory examination of the historical record reveals,
'geometric Brownian motion' is at best a first approximation
to the actual movements of the price of any real stock or collection of stocks.
Even their assumption that the governing processes are stochastic – rather than
examples of deterministic chaos – may in time be disproved
by sufficiently sensitive measurement techniques.

JAMES CASE, reviewing (Mantegna and Stanley, 2000)

The Brownian motion model is extremely popular,
not primarily because of statistical evidence,
but because it is only with this model that we can determine option prices exactly.

ROBERT F. ALMGREN (Almgren, 2002)

As a graduate student at the London School of Economics
I was taught that stock markets were efficient.
Broadly this means that all outstanding information about companies

is built into their share prices, i.e. they are always fairly valued.
This sad fact was hammered home to students with a series of studies
demonstrating that stock-market brokers and analysts,
people with the very best information,
fared no better in their stock-market selections than a monkey
drawing a name from a hat,
or a man throwing darts at the pages of the *Wall Street Journal*.
The first implication of the so-called efficient markets theory is that
there is no sure way to make money in the stock market
other than trading on inside information.
Milken, and others on Wall Street, saw that this simply was not true.
The market, which may have been quick to digest earnings data,
was grossly inefficient in valuing everything from the land a company owns to the pension
fund it creates.

MICHAEL LEWIS (Lewis, 1989)

A trade takes place when the greediest buyer,
afraid that prices will run away from him,
steps up and bids a penny more.
Or the most fearful seller, afraid of getting stuck with his merchandise,
agrees to accept a penny less.

ALEXANDER ELDER (Elder, 2002)

7

Asset price model: Part II

OUTLINE

- computing discrete asset paths
- timescale invariance
- sum-of-square returns

7.1 Computing asset paths

Having derived the model, we may use (6.9) to generate computer simulations of asset prices. Suppose we wish to simulate the evolution of $S(t)$ at certain points $\{t_i\}_{i=0}^{K}$, with $0 = t_0 < t_1 < t_2 < \cdots < t_K = T$. We may compute values $\{S_i\}_{i=0}^{K}$ according to

$$S_{i+1} = S_i e^{(\mu - \frac{1}{2}\sigma^2)(t_{i+1} - t_i) + \sigma \sqrt{t_{i+1} - t_i}\,\xi_i}, \tag{7.1}$$

where each ξ_i is a sample from a $\mathsf{N}(0, 1)$ pseudo-random number generator. The resulting points (t_i, S_i) form a *discrete asset path*.

Computational example Figure 7.1 shows the results of such a simulation with 10^3 time points equally spaced in $[0, 3]$. We took $S_0 = 1$, $\mu = 0.05$ and $\sigma = 0.1$. To produce the picture, we followed the usual convention of joining the discrete data points (t_i, S_i) by straight lines. Overall, the resulting picture agrees qualitatively with typical asset price plots, such as those in Figures 5.1 and 5.2. $\diamond$

To obtain the picture in Figure 7.1 we computed a discrete, but closely spaced, set of data points and joined them with straight lines. The picture seems to suggest that the points lie on a continuous, but 'jagged', curve. This concept can be formalized. On the one hand it can be shown that, with probability 1, an asset path arising from the $\delta t \to 0$ limit in (6.2) will be a continuous function of t. But on the other hand it can also be shown that, with probability 1, the path will not have a well-defined tangent at any point.

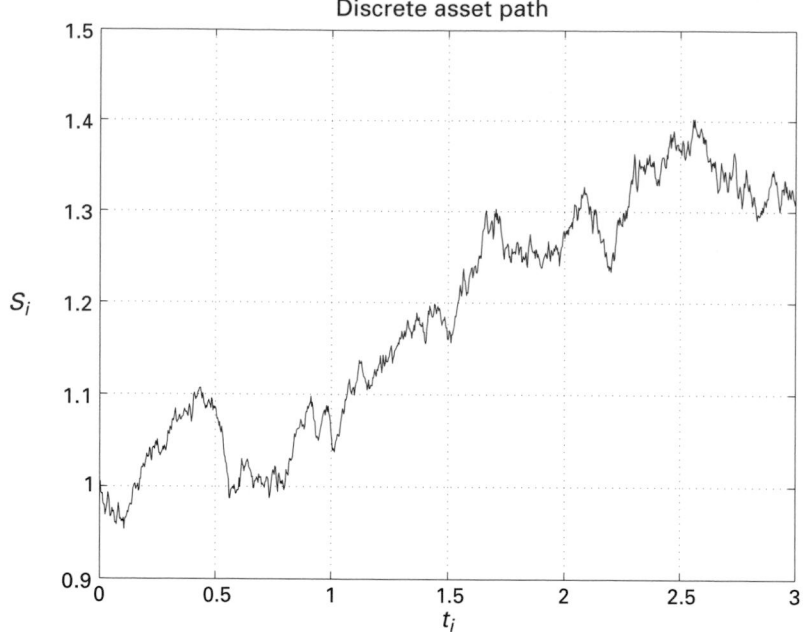

Fig. 7.1.   Discrete asset path of the form (7.1). Discrete points are joined by straight lines to give the impression of a continuous curve.

We would also expect from the original discrete model (6.2) that increasing the volatility parameter σ should turn up the 'jaggedness'. The next computational example tests for this effect.

Computational example   Figure 7.2 shows asset paths computed with the same parameters as for Figure 7.1, except that we set $\sigma = 0.2$ in the upper picture and $\sigma = 0.4$ in the lower picture. The same psuedo-random number sequence $\{\xi_i\}$ was used in both cases. The results confirm that the volatility parameter σ controls the jaggedness of the path.   ◇

Although individual asset paths are nonsmooth functions, we know from (6.11) that the mean of $S(t)$ is smooth. This is confirmed in the next computational example.

Computational example   Here we take $\mu = 0.2$ and $\sigma = 0.3$ and use 10^3 equally spaced time points over $[0, 3]$. We generated 10^4 such discrete paths, starting from $S_0 = 1$ but using different random number generator samples for each path. The upper picture in Figure 7.3 shows the first 20 such paths. In the lower picture we plot the sample mean: at each time point we plot the average of the 10^4 different asset values. We see that this sample mean is indeed smooth;

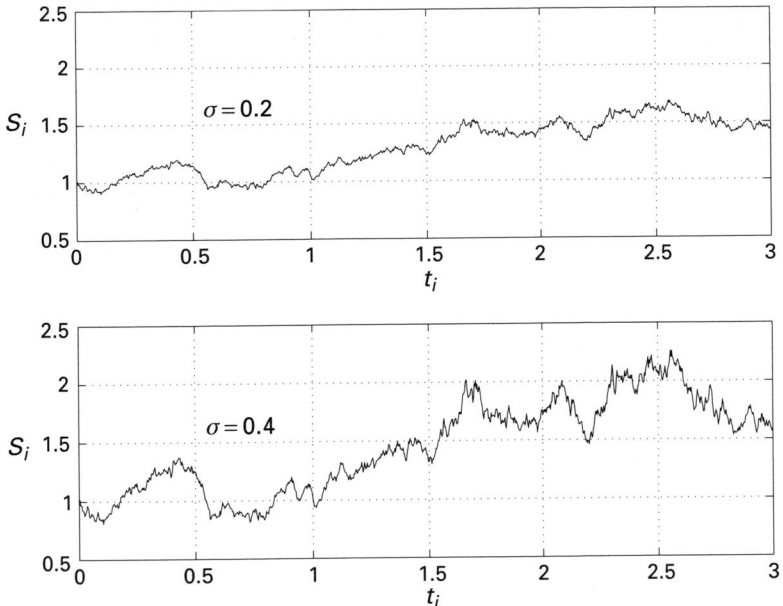

Fig. 7.2. Two discrete asset paths of the form (7.1). Lower picture has higher volatility.

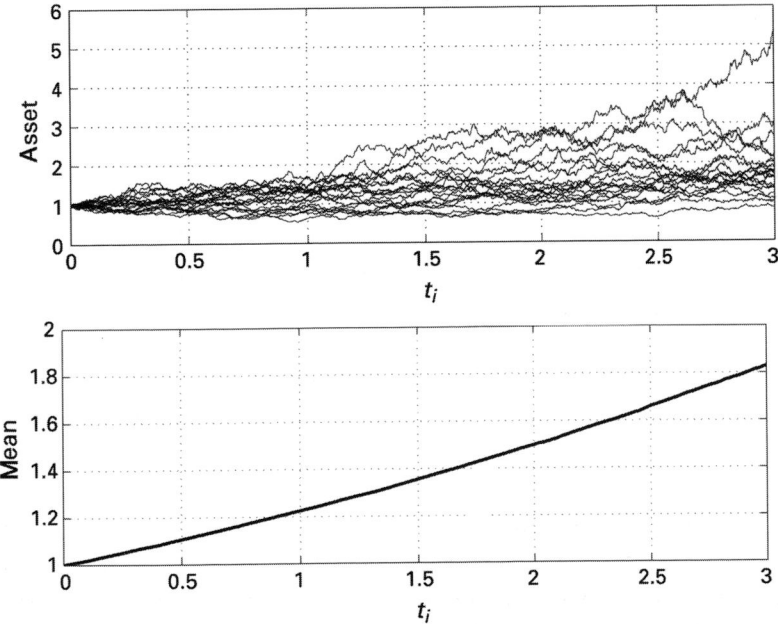

Fig. 7.3. Upper picture: 20 discrete asset paths. Lower picture: sample mean of 10^4 discrete asset paths.

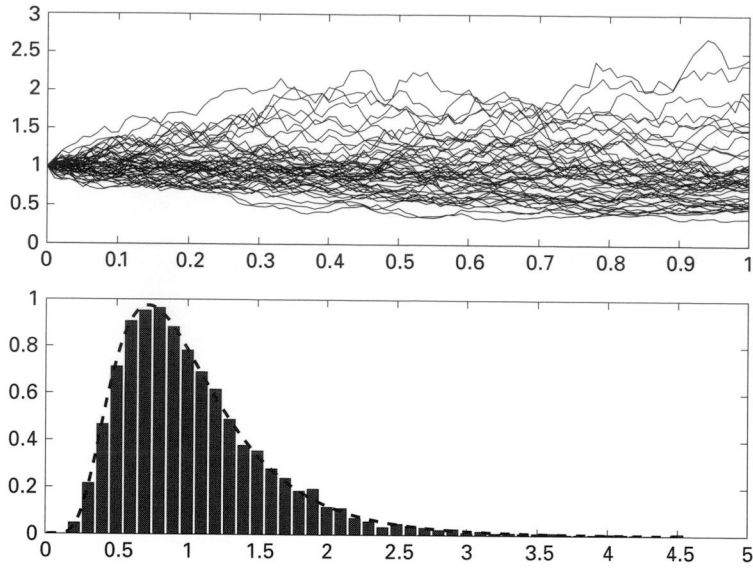

Fig. 7.4.   Upper picture: 50 discrete asset paths over $[0, T]$ with $S_0 = 1$, $\mu = 0.05$, $\sigma = 0.5$, $T = 1$ and $\delta t = 10^{-2}$. Lower picture: histogram for $S(T)$ from 10^4 such paths, with lognormal density function (6.10) superimposed.

it is visually indistinguishable from the exact mean $S_0 e^{\mu t}$ that we derived in (6.11).   ◇

We next give a test that confirms the lognormal behaviour of the asset model.

Computational example   Here, we set $S_0 = 1$, $\mu = 0.05$ and $\sigma = 0.5$, and computed discrete paths over $[0, T]$, with $T = 1$. We used a uniform time spacing of $t_{i+1} - t_i = \delta t = 10^{-2}$. The upper picture in Figure 7.4 shows 50 such paths. In the lower picture we give a kernel density estimate for the asset price at expiry. This was computed in the manner discussed in Section 4.3, using a histogram with 45 bins of width 0.05. The corresponding lognormal density function (6.10), which is superimposed as a dashed line, gives a good match.   ◇

7.2 Timescale invariance

The next computational example reveals a key property of the asset price model. *The jaggedness looks the same over a range of different timescales.* In other words, zooming in or out of the picture, we see the same qualitative behaviour. We saw the same effect when we moved from daily to weekly data in Figures 5.1 and 5.2.

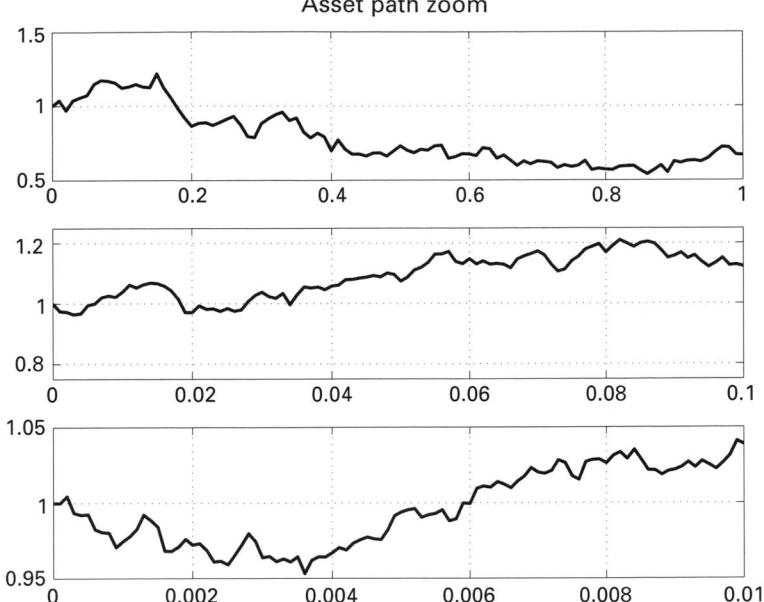

Fig. 7.5. The same asset path sampled at different scales. Upper picture: 100 samples over $[0, 1]$. Middle picture: 100 samples over $[0, 0.1]$. Lower picture: 100 samples over $[0, 0.01]$.

Computational example To generate Figure 7.5, we computed a single asset path for $S_0 = 1$, $\mu = 0.05$ and $\sigma = 0.5$ at equally spaced time points in $[0, 1]$ a distance 10^{-4} apart. Using this data, we plot three pictures. Each picture shows the path at 100 equally spaced time points.

- The upper plot shows the path at 100 equally spaced points in $[0, 1]$.
- The middle plot shows the path at 100 equally spaced points in $[0, 0.1]$.
- The lower plot shows the path at 100 equally spaced points in $[0, 0.01]$

We see that zooming in on the path in this manner does not reveal any change in the qualitative features – the path is 'jagged' at all time scales. $\diamond$

To understand why the pictures have this 'timescale stability' we go back to the discrete model (6.2) and consider

- a small time interval δt,
- very small time interval $\widehat{\delta t} = \delta t / L$, where L is a large integer. (In Figure 7.5 we used quite a moderate value, $L = 10$.)

Using (6.2) to get from time $t = 0$ to $t = \widehat{\delta t}$ we have

$$S(\widehat{\delta t}) - S_0 = S_0(\mu \widehat{\delta t} + \sigma \sqrt{\widehat{\delta t}} Y_0) = S_0 N(\mu \widehat{\delta t}, \sigma^2 \widehat{\delta t}) \qquad (7.2)$$

for the change in $S(t)$. From time $t = 0$ to $t = \delta t$, increments like this add up:

$$S(\delta t) - S_0 = \sum_{i=0}^{L-1} \left(S((i+1)\widehat{\delta t}) - S(i\widehat{\delta t})\right) = \sum_{i=0}^{L-1} S(i\widehat{\delta t})(\mu\widehat{\delta t} + \sigma\sqrt{\widehat{\delta t}}Y_i).$$

Approximating[1] each $S(i\widehat{\delta t})$ by S_0 and using insight from the Central Limit Theorem suggests that

$$S(\delta t) - S_0 \approx S_0 \sum_{i=0}^{L-1} \left(\mu\widehat{\delta t} + \sigma\sqrt{\widehat{\delta t}}Y_i\right) = S_0\mathsf{N}(\mu L\widehat{\delta t}, \sigma^2 L\widehat{\delta t}) = S_0\mathsf{N}(\mu\delta t, \sigma^2\delta t),$$

which reproduces (7.2) over the longer timescale.

7.3 Sum-of-square returns

In Section 5.3 we introduced the concept of the return of an asset; this is simply the relative price change. For small $\delta t = t_{i+1} - t_i$ our original discrete model (6.2) assumes that

$$\frac{S(t_{i+1}) - S(t_i)}{S(t_i)} = \mu\delta t + \sigma\sqrt{\delta t}Y_i, \tag{7.3}$$

so the return is an $\mathsf{N}(\mu\delta t, \sigma^2\delta t)$ random variable. Under this model we know the *statistics* of the return – given any numbers a and b we can work out the probability that the return over the next interval lies between a and b, but, of course, we cannot predict with any certainty what actual return will be seen.

By contrast with the uncertainty of returns, we can show that the *sum-of-square returns* **is** predictable. Suppose the interval $[0, t]$ is divided into a large number of equally spaced subintervals $[0, t_1], [t_1, t_2], \ldots, [t_{L-1}, t_L]$, with $t_i = i\delta t$ and $\delta t = t/L$. Then from (7.3) it is straightforward to show that

$$\mathbb{E}\left[\left(\frac{S(t_{i+1}) - S(t_i)}{S(t_i)}\right)^2\right] = \sigma^2\delta t + \text{higher powers of } \delta t, \tag{7.4}$$

and

$$\text{var}\left[\left(\frac{S(t_{i+1}) - S(t_i)}{S(t_i)}\right)^2\right] = 2\sigma^4\delta t^2 + \text{higher powers of } \delta t, \tag{7.5}$$

see Exercise 7.1.

Hence, using insight from the Central Limit Theorem, $\sum_{i=0}^{L-1}((S(t_{i+1}) - S(t_i))/S(t_i))^2$ should behave like $\mathsf{N}(L\sigma^2\delta t, L2\sigma^4\delta t^2)$, that is, $\mathsf{N}(\sigma^2 t, 2\sigma^4 t\delta t)$. This random variable has a variance proportional to δt, and hence is essentially

---

[1] Some justification for this type of approximation can be found in Section 8.2.

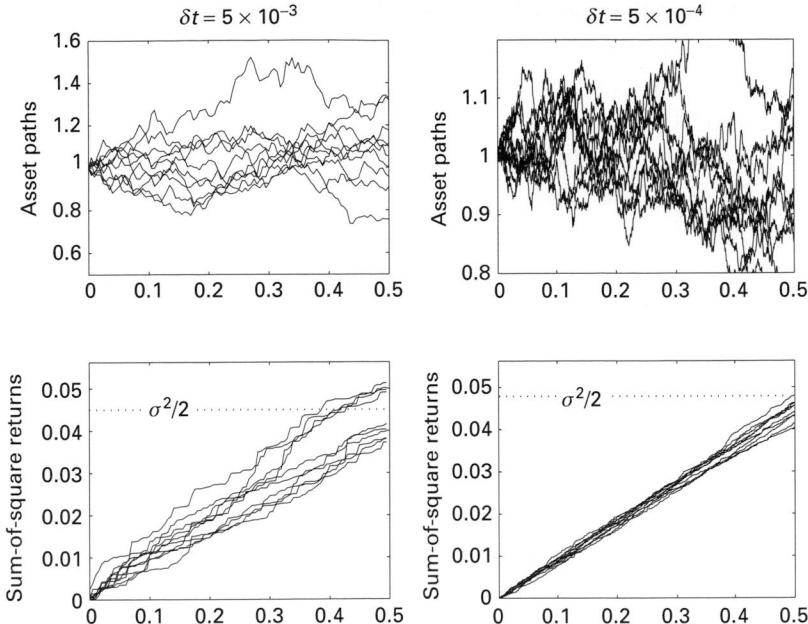

Fig. 7.6. Upper pictures: asset paths. Lower pictures: running sum-of-square returns (7.6).

constant. Thus, although the individual returns are unpredictable, the sum of the squared returns taken over a large number of small intervals is approximately equal to $\sigma^2 t$.

Computational example Figure 7.6 confirms the sum-of-square returns result. We use $S_0 = 1$, $\mu = 0.05$ and $\sigma = 0.3$. Ten asset paths over $[0, 0.5]$ are shown in the upper left plot. The paths were computed using equally spaced time points a distance $\delta t = 0.5/100 = 5 \times 10^{-3}$ apart, so $L = 100$. The lower left picture plots the running sum-of-square returns

$$\sum_{i=1}^{k} \left(\frac{S(t_{i+1}) - S(t_i)}{S(t_i)} \right)^2 \tag{7.6}$$

against t_k for each path. The sum is seen to approximate $\sigma^2 t_k$; the height $\sigma^2/2$ is shown as a dotted line. The right-hand pictures repeat the experiment with $L = 10^3$, so $\delta t = 5 \times 10^{-4}$. We see that reducing δt has improved the match. $\diamond$

7.4 Notes and references

Our treatment of timescale invariance in Section 7.2 can be made rigorous, but the concepts required are beyond the scope of this book. (The essence is that if $W(t)$ is a Brownian motion then so is $W(c^2 t)/c$, for any constant $c > 0$; see, for example, (Brzeźniak and Zastawniak, 1999, Exercise 6.28) and (Brzeźniak and Zastawniak, 1999, Exercise 7.20), and their solutions, for details of this result and why it applies to the asset model.)

There have been numerous attempts to develop generalizations or alternatives to the lognormal asset price model. Many of these are motivated by the observation that real market data has *fat tails* – extreme events occur more frequently than a model based on normal random variables would predict.

One approach is to allow the volatility to be stochastic, see (Duffie, 2001; Hull, 2000; Hull and White, 1987), for example. Another is to allow the asset to undergo 'jumps', see (Duffie, 2001; Hull, 2000; Kwok, 1998), for example. Jump models are especially popular for modelling assets from the utility industries, such as electrical power. The article (Cyganowski *et al.*, 2002) discusses some implementation issues.

An alternative is to take a general, parametrized class of random variables and fit the parameters to stock market data, see (Rogers and Zane, 1999), for example.

A completely different approach is to abandon any attempt to *understand* the processes that drive asset prices (in particular to pay no heed to the efficient market hypothesis) and instead to test as many models as possible on real market data, and use whatever works best as a predictive tool. A group of mathematical physicists with expertise in chaos and nonlinear time series, led by Doyne Farmer and Norman Packard, took up this idea. They founded The Prediction Company in Santa Fe. The company has a website at www.predict.com/html/ introduction.html which makes the claim that

Our technology allows us to build fully automated trading systems which can handle huge amounts of data, react and make decisions based on that data and execute transactions based on those decisions – all in real time. Our science allows us to build accurate and consistent predictive models of markets and the behavior of financial instruments traded in those markets.

The book (Bass, 1999) gives the story behind the foundation and early years of the company and has many insights into the practical issues involved in collecting and analysing vast amounts of financial data.

EXERCISES

7.1. ★ ★ ★ Confirm the results (7.4) and (7.5).

7.2. ★★ By analogy with the continuously compounded interest rate model, we may define the *continuously compounded rate of return* for an asset over $[0, t]$ to be the random variable R satisfying $S(t) = S_0 e^{Rt}$. Using (6.8), show that $R \sim \mathsf{N}(\mu - \sigma^2/2, \sigma^2/t)$.

7.5 Program of Chapter 7 and walkthrough

The program ch07, listed in Figure 7.7, produces a plot of 50 asset paths in the style of the upper picture in Figure 7.4. Having initialized the parameters, we make use of the cumulative product function, cumprod, to produce an array of asset paths. Generally, given an M by L array X, cumprod(X) creates an M by L array whose (i, j) element is the product X(1,j)*X(2,j)*X(3,j)*...*X(i,j). Supplying a second argument set to 2 causes the cumulative product to be taken along the second index – across rows rather than down columns, so cumprod(X,2) creates an M by L array whose (i, j) element is the product X(i,1)*X(i,2)*X(i,3)*...*X(i,j). We also supply two arguments to the randn function: randn(M,L) produces an M by L array with elements from the randn pseudo-random number generator.

It follows that

```
Svals = S*cumprod(exp((mu-0.5*sigma^2)*dt + sigma*sqrt(dt)*randn(M,L)),2);
```

creates an M by L array whose ith row represents a single discrete asset path, as in (6.9). The next line

```
    Svals = [S*ones(M,1) Svals];   % add initial asset price
```

adds the initial asset as a first column, so that the ith row Svals(i,1),Svals(i,2),...,Svals(i,L+1) represents the asset path at times 0,dt,2dt,3dt,...,T.

PROGRAMMING EXERCISES

P7.1. Write a program that illustrates the timescale invariance of the asset model, in the style of Figure 7.5.

P7.2. Use mean and std to verify the approximations (7.4) and (7.5) for (7.3).

```
%CH07   Program for Chapter 7
%
% Plot discrete sample paths

randn('state',100)
clf

%%%%%%%% Problem parameters %%%%%%%%%%%%
S = 1; mu = 0.05; sigma = 0.5; L = 1e2; T = 1; dt = T/L; M = 50;
%%%%%%%%%%%%%%%%%%%%%%%%%%%%%%%%%%%%%%%%

tvals = [0:dt:T];
Svals = S*cumprod(exp((mu-0.5*sigma^2)*dt + sigma*sqrt(dt)*randn(M,L)),2);
Svals = [S*ones(M,1) Svals];  % add initial asset price
plot(tvals,Svals)
title('50 asset paths')
xlabel('t'), ylabel('S(t)')
```

Fig. 7.7.   Program of Chapter 7: ch07.m.

Quotes

But as a warning,
let me note that a trader with a better model might still not be able to transform
this knowledge into money.
Finance is consistent in its ability to build good models
and consistent in its inability to make easy money.
The purpose of the model is to understand the factors
that influence and move option prices
but in the absence of an ability to forecast these factors
the transformation into money remains non-trivial.

DILIP B. MADAN (Madan, 2001)

Evidence countering the efficient market hypothesis
comes in the form of stock market anomalies.
These are events that violate the assumption that stock returns
are randomly distributed.
They include the size effect
(big-company stocks out-perform small-company stocks or vice versa);
the January effect
(stock returns are abnormally high during the first few days of January);
the week-of-the-month effect
(the market goes up at the beginning and down at the end of the month);
and the hour-of-the-day effect
(prices drop during the first hour of trading on Monday and rise on other days).
Prices fall faster than they rise;
the market suffers from 'roundaphobia'
(the Dow breaking ten thousand is a big deal);
and the market tends to overreact
(aggressive buying after good news is followed by nervous selling,
no matter what the news).
Finally, the efficient market hypothesis is incapable of explaining
stock market bubbles and crashes, insider trading, monopolies,
and all the other messy stuff that happens outside its perfect models.

THOMAS A. BASS (Bass, 1999)

Prices reflect intelligent behavior of rational investors and traders,
but they also reflect screaming mass hysteria.

ALEXANDER ELDER (Elder, 2002)

8

Black–Scholes PDE and formulas

8.1 Motivation

At this stage we have defined what we mean by a European call or put option on an underlying asset and we have developed a model for the asset price movement. We are ready to address the key question: what is an option worth? More precisely,

can we systematically determine a *fair* value of the option at $t = 0$?

The answer, of course, is yes, if we agree upon various assumptions. Although our basic aim is to value an option at time $t = 0$ with asset price $S(0) = S_0$, we will look for a function $V(S, t)$ that gives the option value for any asset price $S \geq 0$ at any time $0 \leq t \leq T$. Moreover, we assume that the option may be bought and sold at this value in the market at any time $0 \leq t \leq T$. In this setting, $V(S_0, 0)$ is the required time-zero option value. We are going to assume that such a function $V(S, t)$ exists and is smooth in both variables, in the sense that derivatives with respect to these variables exist. It was mentioned in Section 7.1 that $S(t)$ is **not** a smooth function of t – it is jagged, without a well-defined first derivative. However, it is still perfectly possible for the option value $V(S, t)$ to be smooth in S and t. Looking ahead, Figures 11.3 and 11.4 illustrate this fundamental disparity.

Our analysis will lead us to the celebrated Black–Scholes partial differential equation (PDE) for the function V. The approach is quite general and the PDE is valid in particular for the cases where $V(S, t)$ corresponds to the value of a European call or put.

The key idea in this chapter is *hedging to eliminate risk*. To reinforce the idea, and emphasize that it is a concrete tool as well as a theoretical device, the next chapter is devoted to computational experiments that illustrate hedging in practice.

Before launching into a description of hedging, we first introduce one of the main ingredients that goes into the analysis.

8.2 Sum-of-square increments for asset price

To make progress, we need to work on two timescales. For the rest of the chapter we use

- a *small* timescale, determined by a time increment Δt, and
- a *very small* timescale, determined by a time increment $\delta t = \Delta t / L$, where L is a large integer.

We consider some general time $t \in [0, T]$ and general asset price $S(t) \geq 0$, and focus on the small time interval $[t, t + \Delta t]$. This is broken down into equally spaced, very small, subintervals of length δt, giving $[t_0, t_1], [t_1, t_2], \ldots, [t_{L-1}, t_L]$ with $t_0 = t$, $t_L = t + \Delta t$ and, generally, $t_i = t + i\delta t$.

We will let

$$\delta S_i := S(t_{i+1}) - S(t_i)$$

denote the change in asset price over a very small time increment. Before attempting to derive the Black–Scholes PDE, we need to establish a preliminary result about the sum-of-square increments, $\sum_{i=0}^{L-1} \delta S_i^2$. A similar analysis was done in Section 7.3 for the sum-of-square returns, $\sum_{i=0}^{L-1} (\delta S_i / S(t_i))^2$.

Returning to the discrete model (6.2) we have

$$\delta S_i = S(t_i)(\mu \delta t + \sigma \sqrt{\delta t} Y_i),$$

where the Y_i are i.i.d. $\mathsf{N}(0, 1)$. So

$$\sum_{i=0}^{L-1} \delta S_i^2 = \sum_{i=0}^{L-1} S(t_i)^2 (\mu^2 \delta t^2 + 2\mu\sigma \delta t^{\frac{3}{2}} Y_i + \sigma^2 \delta t Y_i^2). \tag{8.1}$$

We now make this summation amenable to the Central Limit Theorem by replacing each $S(t_i)$ by $S(t)$. This approximation, which is discussed further in the next paragraph, gives us

$$\sum_{i=0}^{L-1} \delta S_i^2 \approx S(t)^2 \sum_{i=0}^{L-1} (\mu^2 \delta t^2 + 2\mu\sigma \delta t^{\frac{3}{2}} Y_i + \sigma^2 \delta t Y_i^2). \tag{8.2}$$

Working out the mean and variance of the random variables inside the summation and appealing to the Central Limit Theorem suggests the approximate relation

$$\sum_{i=0}^{L-1} \delta S_i^2 \sim S(t)^2 N(\sigma^2 L \delta t, 2\sigma^4 L \delta t^2) = S(t)^2 N(\sigma^2 \Delta t, 2\sigma^4 \Delta t \delta t), \qquad (8.3)$$

see Exercise 8.1. Because δt is very small, the variance of that final expression is tiny, leading us to conclude that the sum-of-square increments is approximately a constant multiple of $S(t)^2$:

$$\sum_{i=0}^{L-1} \delta S_i^2 \approx S(t)^2 \sigma^2 \Delta t. \qquad (8.4)$$

The step of replacing each $S(t_i)$ in (8.1) by $S(t)$ can be loosely justified as follows. Our model (6.9) shows that

$$S(t_i) = S(t)e^{(\mu - \frac{1}{2}\sigma^2)i\delta t + \sigma \sqrt{i\delta t}Z}, \qquad \text{for some } Z \sim N(0, 1).$$

Using $e^x \approx 1 + x$ for small x, we have

$$S(t_i) \approx S(t)(1 + \sigma \sqrt{i\delta t}Z)$$

and since $i\delta t \le L\delta t = \Delta t$, we may write, loosely,

$$S(t_i) - S(t) = O(\sqrt{\Delta t}).$$

In words, approximating each $S(t_i)$ by $S(t)$ introduces an error that is roughly proportional to $\sqrt{\Delta t}$. We may thus argue that replacing each $S(t_i)$ in (8.1) with $S(t)$ will not affect the leading term in the approximation (8.4). This is far from a rigorous argument – Z is a random variable, not simply a real number – but it can be shown that the overall conclusion is valid.

Computational example Although we are not in a position to prove (8.4) rigorously, we can certainly illustrate the result via a computational experiment. We may copy the way that Figure 7.6 was produced, but now computing the sum-of-square increments, instead of the sum-of-square returns. We set $S_0 = 1$, $\mu = 0.05$ and $\sigma = 0.3$. The upper left plot in Figure 8.1 shows ten discrete asset paths over $[0, \Delta t]$ with $\Delta t = 0.5$, using equally spaced points a distance $\delta t = \Delta t/100 = 5 \times 10^{-3}$ apart. So $L = 100$ and $t = 0$. The lower left picture plots the running sum-of-square increments

$$\sum_{i=1}^{k} \delta S_i^2 \qquad (8.5)$$

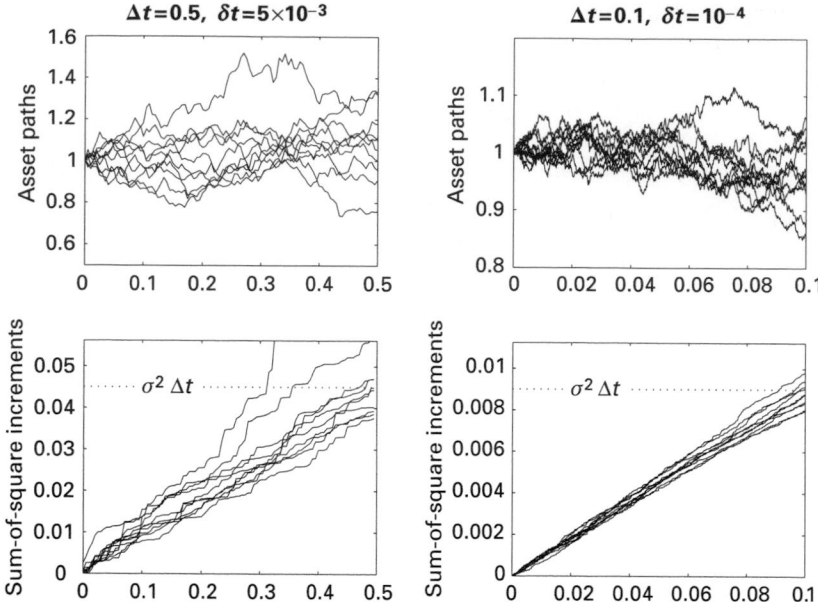

Fig. 8.1.   Upper pictures: asset paths. Lower pictures: running sum-of-square increments (8.5).

against t_k for each path. We see that the sum typically approximates $\sigma^2 \Delta t = 0.045$ as k approaches L. The right-hand pictures give the same information for an example with $\Delta t = 0.1$ and $L = 1000$, so $\delta t = 10^{-4}$. We see that the quality of the approximation (8.4) has improved.   ◇

8.3 Hedging

Now, to find a fair option value, we set up a *replicating portfolio* of asset and cash, that is, a combination of asset and cash that has *precisely the same risk* as the option at all time. The portfolio will consist of a cash deposit D and a number A of units of asset. We allow D and A to be functions of asset price S and time t. The portfolio value, denoted by Π, thus satisfies

$$\Pi(S, t) = A(S, t)S + D(S, t). \tag{8.6}$$

We must specify how the asset holding $A(S, t)$ and cash deposit $D(S, t)$ are going to vary with S and t. Before delving into the details it is perhaps useful to remind ourselves of some basic assumptions that are being made, all of which have been introduced earlier:

- there are no transaction costs,
- the asset can be bought/sold in arbitrary units,

- short selling is permitted,
- no dividends are paid,
- the interest rate r is constant,
- trading of the asset (and option) can take place in continuous time.

To avoid unreadably long equations we will also introduce some shorthand notation. A subscript i denotes evaluation of a function at $(S(t_i), t_i)$, so

$$V_i \text{ means } V(S(t_i), t_i), \ \Pi_i \text{ means } \Pi(S(t_i), t_i), \text{ etc.}$$

No subscript denotes evaluation at $(S(t), t)$, so

$$V \text{ means } V(S(t), t), \ \Pi \text{ means } \Pi(S(t), t), \text{ etc.}$$

The symbol δ denotes the difference over a timestep of length δt, so

- δS_i means $S(t_{i+1}) - S(t_i)$,
- δV_i means $V(S(t_{i+1}), t_{i+1}) - V(S(t_i), t_i)$,
- $\delta \Pi_i$ means $\Pi(S(t_{i+1}), t_{i+1}) - \Pi(S(t_i), t_i)$,
- $\delta(V - \Pi)_i$ means $\delta V_i - \delta \Pi_i$, etc.

Our strategy for the portfolio (8.6) is to keep the amount of asset constant over each very small timestep of length δt. It follows that the change in the value of the portfolio has two sources.

(1) The asset price fluctuation. The change δS_i produces a change $A_i \delta S_i$ in the portfolio value.
(2) Interest accrued on the cash deposit. Using the discrete version for convenience (see (2.7) in Exercise 2.2), we may write this contribution to the portfolio change as $r D_i \delta t$.

Overall,

$$\delta \Pi_i = A_i \delta S_i + r D_i \delta t. \tag{8.7}$$

Now because V is assumed to be a smooth function of S and t, a Taylor series expansion gives

$$\delta V_i \approx \frac{\partial V_i}{\partial t} \delta t + \frac{\partial V_i}{\partial S} \delta S_i + \tfrac{1}{2} \frac{\partial^2 V_i}{\partial S^2} \delta S_i^2. \tag{8.8}$$

We have kept the δS_i^2 term in (8.8) because experience from the previous two chapters suggests that it will make a contribution of size proportional to δt. Subtracting (8.7) from (8.8) in order to compare the change in the portfolio with that in the option value, we find

$$\delta(V - \Pi)_i \approx \left(\frac{\partial V_i}{\partial t} - r D_i \right) \delta t + \left(\frac{\partial V_i}{\partial S} - A_i \right) \delta S_i + \tfrac{1}{2} \frac{\partial^2 V_i}{\partial S^2} \delta S_i^2. \tag{8.9}$$

Our aim is to make the portfolio replicate the option, so that the difference between them is predictable. We can eliminate the unpredictable δS_i term from (8.9) by setting

$$A_i = \frac{\partial V_i}{\partial S},$$ (8.10)

in which case

$$\delta(V - \Pi)_i \approx \left(\frac{\partial V_i}{\partial t} - rD_i\right)\delta t + \frac{1}{2}\frac{\partial^2 V_i}{\partial S^2}\delta S_i^2.$$ (8.11)

The final step in eliminating randomness is to add these differences over $0 \leq i \leq L - 1$ and exploit (8.4), which shows that the sum of the δS_i^2 terms is nonrandom.

Before proceeding with that final step, we pause to explain what (8.10) means in practice. If we are able to find the required function V, then we may differentiate it with respect to S in order to specify our strategy for updating the portfolio. At the end of the step from t_i to t_{i+1} we rebalance our asset holding to $A_{i+1} = \partial V_{i+1}/\partial S$. This may involve selling (if $\partial V_{i+1}/\partial S < \partial V_i/\partial S$) or buying (if $\partial A_{i+1}/\partial S > \partial A_i/\partial S$) some amount of the asset. We want to make the portfolio *self-financing*, that is, beyond time $t = 0$ we do not want to add or remove money. This can be achieved by using the cash account to finance the update – the money needed for, or generated by, the asset rebalancing, is reflected by a corresponding change from D_i to D_{i+1}. This idea of continually fine-tuning the portfolio in order to reduce or remove risk is known as *hedging*.

8.4 Black–Scholes PDE

Letting $\Delta(V - \Pi)$ denote the change in $V - \Pi$ from time t to $t + \Delta t$, that is,

$$\Delta(V - \Pi) = V(S(t + \Delta t), t + \Delta t) - \Pi(S(t + \Delta t), t + \Delta t)$$
$$- (V(S(t), t) - \Pi(S(t), t)),$$

we may sum (8.11) to give

$$\Delta(V - \Pi) \approx \sum_{i=0}^{L-1}\left(\frac{\partial V_i}{\partial t} - rD_i\right)\delta t + \frac{1}{2}\sum_{i=0}^{L-1}\frac{\partial^2 V_i}{\partial S^2}\delta S_i^2.$$ (8.12)

On the basis that V and D are smooth functions, we will replace the arguments $S(t_i), t_i$ in $\partial V_i/\partial t$, D_i and $\partial^2 V_i/\partial S^2$, by $S(t), t$, in a similar manner to the approximation used for (8.1). So, using $L\delta t = \Delta t$,

$$\Delta(V - \Pi) \approx \left(\frac{\partial V}{\partial t} - rD\right)\Delta t + \frac{1}{2}\frac{\partial^2 V}{\partial S^2}\sum_{i=0}^{L-1}\delta S_i^2.$$

Now, using (8.4), and assuming that all approximations are exact in the limit $\delta t \to 0$, we may write

$$\Delta(V - \Pi) = \left(\frac{\partial V}{\partial t} - rD + \tfrac{1}{2}\sigma^2 S^2 \frac{\partial^2 V}{\partial S^2} \right) \Delta t. \qquad (8.13)$$

The final leap of logic is to argue that because this change in the portfolio $V - \Pi$ is nonrandom, it must equal the corresponding growth offered by the risk-free interest rate, so

$$\Delta(V - \Pi) = r \, \Delta t (V - \Pi). \qquad (8.14)$$

This follows from the no arbitrage principle. If $\Delta(V - \Pi) > r \, \Delta t (V - \Pi)$ then we could make a guaranteed profit greater than that offered by the risk-free interest rate by

(i) acquiring the portfolio $V - \Pi$ at time t – buying the option at V in the marketplace, and selling the portfolio Π (i.e. short selling A units of asset and loaning out an amount D of cash), and
(ii) selling the portfolio $V - \Pi$ at time $t + \Delta t$.

Similarly, if $\Delta(V - \Pi) < r \, \Delta t (V - \Pi)$ then we could make a guaranteed profit greater than that offered by the risk-free interest rate by

(i) selling the portfolio $V - \Pi$ at time t – selling the option at V in the marketplace, and buying the portfolio Π (i.e. buying A units of asset and borrowing an amount D of cash), and
(ii) buying the portfolio $V - \Pi$ at time $t + \Delta t$.

Now, combining (8.6), (8.13) and (8.14) gives

$$\frac{\partial V}{\partial t} - rD + \tfrac{1}{2}\sigma^2 S^2 \frac{\partial^2 V}{\partial S^2} = r(V - AS - D).$$

Using $A = \partial V / \partial S$ from (8.10) and rearranging, we arrive at

$$\frac{\partial V}{\partial t} + \tfrac{1}{2}\sigma^2 S^2 \frac{\partial^2 V}{\partial S^2} + rS \frac{\partial V}{\partial S} - rV = 0. \qquad (8.15)$$

This is the famous *Black–Scholes* partial differential equation (PDE). It is a relationship between V, S, t and certain partial derivatives of V.

Two points are worth raising immediately.

(1) The drift parameter μ in the asset model does not appear in the PDE.
(2) We have not yet specified what type of option is being valued. The PDE must be satisfied for **any** option on S whose value can be expressed as some smooth function $V(S, t)$.

Regarding point (2), to determine $V(S, t)$ uniquely we must specify other conditions that involve information about the particular option. As is typical with many differential equations, these will apply somewhere along the edges of the domain $0 \leq S, 0 \leq t \leq T$ on which the problem is posed.

We will use $C(S, t)$ to denote the European call option value. In this case, we know for certain that at the expiry time, $t = T$, the payoff is $\max(S(T) - E, 0)$. This **must** be the value of the option at time T, otherwise an obvious arbitrage opportunity exists. So

$$C(S, T) = \max(S(T) - E, 0). \qquad (8.16)$$

Now if the asset price is ever zero, then it is clear from (6.9) that $S(t)$ remains zero for all time and hence the payoff will be zero at expiry. So, in this case, the value of the option must be zero at all times. Hence,

$$C(0, t) = 0, \quad \text{for all } 0 \leq t \leq T. \qquad (8.17)$$

Conversely, if the asset price is ever extremely large, then it is very likely to remain extremely large and swamp the exercise price, so that,

$$C(S, t) \approx S, \quad \text{for large } S. \qquad (8.18)$$

The constraint (8.16) is called a *final condition*, as it applies at the final time $t = T$. It is much more common to come across initial conditions, specified at $t = 0$, and we will see in Chapter 24 that the PDE is easily transformed into such a problem. The other constraints, (8.17) and (8.18), are known as *boundary conditions*.

8.5 Black–Scholes formulas

Imposing (8.16), (8.17) and (8.18) on the Black–Scholes PDE (8.15) is enough to force a unique solution to exist for the call option value. (In fact we could get away with less boundary information, see Section 8.6.) This solution is

$$C(S, t) = SN(d_1) - Ee^{-r(T-t)}N(d_2), \qquad (8.19)$$

where $N(\cdot)$ is the N(0, 1) distribution function, defined in (3.18), and

$$d_1 = \frac{\log(S/E) + (r + \frac{1}{2}\sigma^2)(T - t)}{\sigma\sqrt{T - t}}, \qquad (8.20)$$

$$d_2 = \frac{\log(S/E) + (r - \frac{1}{2}\sigma^2)(T - t)}{\sigma\sqrt{T - t}}. \qquad (8.21)$$

We may also write

$$d_2 = d_1 - \sigma \sqrt{T - t}, \tag{8.22}$$

see Exercise 8.2. The equation (8.19) displays the *Black–Scholes formula* for the value of a European call. It is possible to construct the formula by solving the PDE (8.15) under (8.16), (8.17) and (8.18). In this book, we take the easier route of verifying directly that $C(S, t)$ in (8.19) has the right properties. Exercise 8.3 deals with (8.16), (8.17) and (8.18), and Section 10.4 deals with the PDE (8.15).

Having obtained a formula for a European call option value, we may exploit put–call parity to establish the value $P(S, t)$ of a European put option. In Section 2.5 we derived the relation (2.2) that connects the time-zero call and put values. Letting $P(S, t)$ denote the put value at asset price S and time t, the same argument gives the general put–call parity relation

$$C(S, t) + Ee^{-r(T-t)} = P(S, t) + S, \tag{8.23}$$

see Exercise 8.4. Combining (8.19) and (8.23) leads to the Black–Scholes formula for the value of a European put option,

$$P(S, t) = Ee^{-r(T-t)} (1 - N(d_2)) + S (N(d_1) - 1).$$

Using Exercise 3.9, this may be simplified to

$$P(S, t) = Ee^{-r(T-t)} N(-d_2) - SN(-d_1). \tag{8.24}$$

Alternatively, we could derive final time and boundary conditions and attempt to solve the Black–Scholes PDE. Since the payoff for a put option at time $t = T$ is $\max(E - S(T), 0)$, we have

$$P(S, T) = \max(E - S(T), 0). \tag{8.25}$$

If the asset price is ever zero then $S(T) = 0$ and the payoff at time T will be E. To obtain $P(0, t)$ we discount for inflation, to get

$$P(0, t) = Ee^{-r(T-t)}, \quad \text{for all } 0 \leq t \leq T. \tag{8.26}$$

For extremely large S the payoff is almost certain to be zero, so

$$P(S, t) \approx 0, \quad \text{for large } S. \tag{8.27}$$

Exercise 8.5 asks you to confirm that $P(S, t)$ in (8.24) satisfies the conditions (8.25)–(8.27) and in Exercise 10.7 of Chapter 10 you are set the task of showing that it solves the Black–Scholes PDE (8.15).

Computational example For illustration, we give a simple example of evaluating the Black–Scholes formulas. With $t = 0$, $S_0 = 5$, $E = 4$, $T = 1$, $\sigma = 0.3$

and $r = 0.05$, we find, to four decimal places,

$$d_1 = 1.0605,$$
$$d_2 = 0.7605,$$
$$N(d_1) = 0.8555,$$
$$N(d_2) = 0.7765,$$
$$N(-d_1) = 0.1445,$$
$$N(-d_2) = 0.2235.$$

Here, we used MATLAB's `erf` function in order to evaluate $N(x)$ – see Exercise 4.1. The resulting European call and put option values are

$$C(5, 0) = 1.3231 \quad \text{and} \quad P(5, 0) = 0.1280.$$

The put–call parity relation (2.2) is easily confirmed.   ◇

8.6 Notes and references

The two classic references for the Black–Scholes theory are the paper (Black and Scholes, 1973) by Fischer Black and Myron S. Scholes, which derives the key equations, and the paper (Merton, 1973) by Robert C. Merton, which adds a rigorous mathematical analysis. Merton and Scholes were awarded the 1997 Nobel Prize in Economic Sciences for this work. It is widely accepted that Fischer Black, who died in 1995, would have shared in the prize had he still been alive. Details of the prize can be found at www.nobel.se/economics/laureates/1997/.
The accompanying press release argues that

> A new method to determine the value of derivatives stands out among the foremost contributions to economic sciences over the last 25 years.

The heuristic, discrete-time treatment of hedging that we used to derive the Black–Scholes PDE was inspired by the expository article of Almgren (Almgren, 2002). Modern texts that give rigorous derivations of the Black–Scholes formula include (Björk, 1998; Duffie, 2001; Karatzas and Shreve, 1998; Nielsen, 1999; Øksendal, 1998).

It is possible to weaken the boundary conditions (8.17) and (8.18) in the Black–Scholes PDE (8.15) without sacrificing uniqueness of the solution. Some control on the growth of the solution as $S \to \infty$ would suffice, see for example (Wilmott *et al.*, 1995). We will return to the issue of boundary conditions when we discuss finite difference methods in Chapters 23 and 24.

As a final comment, we note that although the time-T call value is a nonsmooth hockey stick, (8.16), the function $C(S, t)$ is smooth at all times $0 \leq t < T$; this phenomenon of 'instant smoothing' is typical of diffusion PDEs like (8.15).

EXERCISES

8.1. $\star\star$ Show that (8.2) leads to the approximate relation (8.3). [Hint: use Exercise 3.7.]

8.2. $\star$ Show that (8.21) can be replaced by (8.22).

8.3. $\star\star\star$ Confirm that $C(S, t)$ in (8.19) satisfies (8.16), (8.17) and (8.18). [Hint: to deal with (8.16), take the limit $t \to T^-$, to deal with (8.17) take the limit $S \to 0^+$ and to deal with (8.18) take the limit $S \to \infty$.]

8.4. $\star\star$ Use the argument in Section 2.5 to obtain the general put–call parity relation (8.23).

8.5. $\star\star\star$ Confirm that $P(S, t)$ in (8.24) satisfies (8.25)–(8.27).

8.6. $\star\star$ It is intuitively obvious that call and put options are linear – the value of two options is twice the value of one option. Show how this follows from the Black–Scholes formulas (8.19) and (8.24).

8.7. $\star\star\star$ Show that $\lim_{E \to 0} C(S, t) = S$ in (8.19) and $\lim_{E \to 0} P(S, t) = 0$ in (8.24), and give a financial interpretation of the results.

8.8. $\star\star\star$ Write down a PDE and final time/boundary conditions for the value of a butterfly spread, as described in Exercise 1.3.

8.9. $\star\star\star$ Verify that

$$V(S, t) = \frac{e^{(\sigma^2 - 2r)(T - t)}}{S}$$

is a solution of the Black–Scholes PDE (8.15). What is the practical implication of this result?

8.10. $\star\star\star$ Verify that S and e^{rt} are solutions of the Black–Scholes PDE (8.15) and give an accompanying financial explanation.

8.11. $\star\star\star$ Consider the problem posed in Exercise 2.6 of finding a fair value for a forward contract. Use Exercise 8.7 above to confirm that $F = S(0)e^{rT}$.

8.7 Program of Chapter 8 and walkthrough

Unlike the previous seven cases, our code for this chapter, which is listed in Figure 8.2, is a MATLAB *function*. This means that it must be supplied with *input arguments* and it will return *output arguments*. The input arguments S,E,r,sigma and tau represent, respectively, the asset price at time t, the exercise price, the interest rate, the volatility and the time to expiry, $T - t$. It is assumed that tau is non-negative.

```
function [C, Cdelta, P, Pdelta] = ch08(S,E,r,sigma,tau)
% Program for Chapter 8
% This is a MATLAB function
%
% Input arguments: S = asset price at time t
%              E = Exercise price
%              r = interest rate
%              sigma = volatility
%              tau = time to expiry (T-t)
%
% Output arguments: C = call value, Cdelta = delta value of call
%                   P = Put value, Pdelta = delta value of put
%
%   function [C, Cdelta, P, Pdelta] = ch08(S,E,r,sigma,tau)

if tau > 0
  d1 = (log(S/E) + (r + 0.5*sigma^2)*(tau))/(sigma*sqrt(tau));
  d2 = d1 - sigma*sqrt(tau);
  N1 = 0.5*(1+erf(d1/sqrt(2)));
  N2 = 0.5*(1+erf(d2/sqrt(2)));
  C = S*N1-E*exp(-r*(tau))*N2;
  Cdelta = N1;
  P = C + E*exp(-r*tau) - S;
  Pdelta = Cdelta - 1;
else
  C = max(S-E,0);
  Cdelta =  0.5*(sign(S-E) + 1);
  P = max(E-S,0);
  Pdelta = Cdelta - 1;
end
```

Fig. 8.2.   Program of Chapter 8: ch08.m.

The output arguments C,Cdelta,P and Pdelta represent, respectively, the European call, call delta, put and put delta values.

The lines of code between if tau > 0 and else are executed in the case where tau, the time to expiry, is positive. In this case we are evaluating the Black–Scholes values given by (8.19), (8.24), and also the deltas (9.1) and (9.2) that are introduced in Chapter 9, using erf as a means to obtain $N(x)$, as described in Exercise 4.1.

The lines of code between else and end are executed in the remaining case, where tau is zero. Here, we are at expiry and to avoid *division by zero* errors in (8.20) and (8.22), we revert to the expressions (8.16), (8.25), along with (9.7) and (9.8) from Chapter 9. We make use of the signum function, sign, which is defined by

$$\mathrm{sign}(x) = \begin{cases} 1, & \text{if } x > 0, \\ 0, & \text{if } x = 0, \\ -1, & \text{if } x < 0. \end{cases}$$

An example of the function in use is

```
>> S = 2; E = 2.5; r = 0.03; sigma = 0.25; tau = 1;
>> [C, Cdelta, P, Pdelta] = ch08(S,E,r,sigma,tau)
```

which outputs

```
C = 0.0691
Cdelta = 0.2586
P = 0.4953
Pdelta = -0.7414
```

PROGRAMMING EXERCISES

P8.1. Use ch08.m to produce graphs illustrating the limits $\lim_{t \to T^-} C(S, t) = \max(S(T) - E, 0)$ and $\lim_{S \to \infty} C(S, t) = S$ established in Exercise 8.3.

P8.2. Write a program that illustrates (8.4) in the style of Figure 8.1.

Quotes

Stephen Belloti: 'Myron, what do you have more of – money or brains?'
Myron Scholes: 'Brains, but it's getting close.'

Source (Lowenstein, 2001)

In the early 1970s, Merton tackled a problem
that had been partially solved by two other economists,
Fischer Black and Myron S. Scholes:
deriving a formula for the 'correct' price of a stock option.
Grasping the intimate relation between an option and the underlying stock,
Merton completed the puzzle with an elegantly mathematical flourish.
Then he graciously waited to publish until after his peers did;
thus the formula would ever be known as the Black–Scholes model.
Few people would have cared given that no active market for options existed.
But coincidentally, a month before the formula appeared,
the Chicago Board Options Exchange had begun to list stock options for trading.
Soon, Texas Instruments was advertising in *The Wall Street Journal*,
'Now you can find the Black–Scholes value using our ... calculator.'
This was the true beginning of the derivatives revolution.
Never before had professors made such an impact on Wall Street.

ROGER LOWENSTEIN (Lowenstein, 2001)

In 1975 I crammed the Black–Scholes formula into a TI-52 handheld calculator,
which was capable of giving me one option price in about thirteen seconds.
It was pretty crude,
but in the land of the blind I was the guy with one eye.

JOE RITCHIE, *option trader, source* (Bass, 1999)

To someone who came out of graduate school in the mid-eighties,
the decade spanning roughly 1969–79
seems like a golden age of dynamic asset pricing theory ...
The Black–Scholes model now seems to be, by far,
the most important single breakthrough of this 'golden decade' ...
Theoretical developments in the period since 1979, with relatively few exceptions,
have been a mopping-up operation.

DARRELL DUFFIE (Duffie, 2001)

9

More on hedging

9.1 Motivation

The hedging idea that was used to derive the Black–Scholes PDE forms the most important concept in this book. In this chapter, we therefore take time out to re-iterate the steps involved and develop the process into an algorithm that can be illustrated numerically.

9.2 Discrete hedging

Having found the explicit formulas (8.19) and (8.24), we may differentiate with respect to S to obtain the required asset holding A_i in (8.10). This partial derivative $\partial V/\partial S$ is called the *delta* of an option, and the hedging strategy that we discussed is known as *delta hedging*. Performing the differentiation leads to

$$\frac{\partial C}{\partial S} = N(d_1) \qquad \text{(delta of a European call)}, \tag{9.1}$$

and

$$\frac{\partial P}{\partial S} = N(d_1) - 1 \qquad \text{(delta of a European put)}. \tag{9.2}$$

Confirmation of these expressions is deferred until Chapter 10, where various partial derivatives are computed.

Returning to the delta hedging process, we know from (8.7) that Π_{i+1}, the value of the portfolio at $t_i + \delta t$, satisfies

$$\Pi_{i+1} = A_i S_{i+1} + (1 + r\delta t)D_i. \tag{9.3}$$

The asset holding is rebalanced to A_{i+1} and in order to compensate, the cash account is altered to D_{i+1}. Since no money enters or leaves the system, the new portfolio value, $A_{i+1}S_{i+1} + D_{i+1}$, must equal Π_{i+1} in (9.3), so

$$D_{i+1} = (1 + r\delta t)D_i + (A_i - A_{i+1})S_{i+1}. \tag{9.4}$$

We may summarize the overall hedging strategy as follows.

> Set $A_0 = \partial V_0 / \partial S$, $D_0 = 1$ (arbitrary), $\Pi_0 = A_0 S_0 + D_0$
> For each new time $t = (i + 1)\delta t$
> > Observe new asset price S_{i+1}
> > Compute new portfolio value Π_{i+1} in (9.3)
> > Compute $A_{i+1} = \frac{\partial V_{i+1}}{\partial S}$
> > Compute new cash holding D_{i+1} in (9.4)
> > New portfolio value is $A_{i+1}S_{i+1} + D_{i+1}$
> end

More precisely, this strategy is *discrete hedging* as the rebalancing act is done at times $i\delta t$. Because we cannot let $\delta t \to 0$ in practice, there will be some error in the risk elimination.

For the purpose of illustration, it is possible to simulate an asset path and implement discrete hedging. To write down the resulting algorithm, we use $\{\xi_i\}$ to denote samples from an $N(0, 1)$ pseudo-random number generator that are used in simulating the asset path, and we let $\delta t = T/N$.

> Set $A_0 = \partial V_0 / \partial S$, $D_0 = 1$ (arbitrary), $\Pi_0 = A_0 S_0 + D_0$
> For $i = 0$ to $N - 1$
> > Compute $S_{i+1} = S_i e^{(\mu - \frac{1}{2}\sigma^2)\delta t + \sqrt{\delta t}\sigma\xi_i}$
> > Set $\Pi_{i+1} = A_i S_{i+1} + (1 + r\delta t)D_i$
> > Compute $A_{i+1} = \frac{\partial V_{i+1}}{\partial S}$
> > Set $D_{i+1} = (1 + r\delta t)D_i + (A_i - A_{i+1})S_{i+1}$
> end

To describe the next set of experiments, it is convenient to use some financial jargon. At time t, a European call option is said to be

> **in-the-money** if $S(t) > E$,
> **out-of-the-money** if $S(t) < E$, and
> **at-the-money** if $S(t) = E$.

The jargon extends in an obvious fashion to other options. In general, in-the-money means that there will be a positive payoff if the asset price stays as it is. Out-of-the-money means that the asset must change by some non-negligible amount in order for a positive payoff to ensue. At-the-money defines the boundary between in- and out-of-the-money.

Computational example Here we implement the discrete hedging simulation above for a European call option with $S_0 = 1$, $E = 1.5$, $\mu = 0.055$, $r = 0.05$, $T = 5$ and $\delta t = 10^{-2}$, so $N = 500$. The upper plot in Figure 9.1 displays the particular discrete asset path (t_i, S_i), for $t_i = i\delta t$, that arose. The strike price E is shown as a dashed line. We see that for this particular asset path, the call option stays out-of-the-money (asset price below E) until just after $t = 1$, and then makes a number of excursions in/out-of-the-money before giving a very small payoff at expiry. The upper-middle plot shows the deltas, $(t_i, \partial C_i / \partial S)$, along the asset path. This shows the time-varying amount of asset held in the portfolio. The lower-middle plot gives the cash level (t_i, D_i) and the solid curve in the lower plot gives the portfolio value (t_i, Π_i). The idea behind delta hedging is to guarantee that the portfolio $C - \Pi$ grows at the risk-free interest rate. It follows that

$$\Pi(S(t), t) = C(S(t), t) - (C(S_0, 0) - \Pi(S_0, 0)) e^{rt} \qquad (9.5)$$

should hold. To test this, we computed the right-hand side of (9.5) at each time t_i, using the Black–Scholes formula (8.19) to compute $C(S_i, t_i)$. Every tenth value has been plotted as a circle in the lower picture.[1] The circles appear to lie on top of the Π_i curve, so (9.5) is approximated well. The discrepancy in (9.5) at the expiry date,

$$\left| C(S(T), T) - \Pi(S(T), T) - (C(S_0, 0) - \Pi(S_0, 0)) e^{rT} \right|, \qquad (9.6)$$

was found to be 0.0364. Reducing δt to 10^{-4} (and hence computing a different asset path), we found that this discrepancy was lowered to 0.0029. ◇

Computational example In Figure 9.2 we repeat the computation in Figure 9.1 with E set to the value 2.5. In this case the option finishes out-of-the-money. Again we observe from the lower picture that (9.5) is close to being exact. ◇

9.3 Delta at expiry

Looking carefully at Figures 9.1 and 9.2 we see that

- in the first experiment, where the option expires in-the-money, the delta approaches the value 1 at expiry, whereas

---

[1] Plotting every value would make the picture too cluttered.

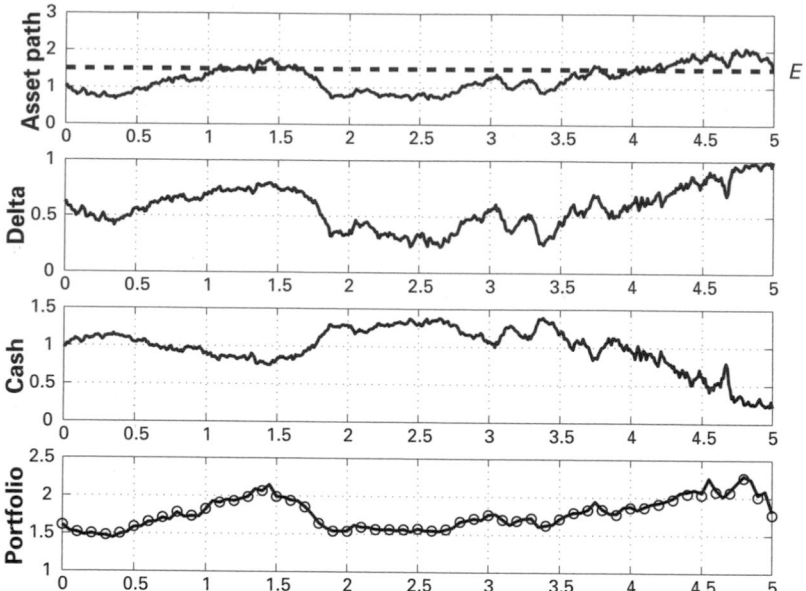

Fig. 9.1.   Discrete hedging simulation. Option expires in-the-money. Upper: discrete asset path. Upper-middle: delta values (also asset holding in portfolio). Lower-middle: cash holding in portfolio. Lower: portfolio value (solid), theoretical portfolio value (9.5) (circles).

- in the second experiment, where the option expires out-of-the-money, the delta approaches the value 0 at expiry.

This is no accident. Using the characterization (9.1), some analysis shows that

$$\lim_{t \to T^-} \frac{\partial C(S,t)}{\partial S} = \begin{cases} 1, & \text{if } S(T) > E, \\ \frac{1}{2}, & \text{if } S(T) = E, \\ 0, & \text{if } S(T) < E, \end{cases} \qquad (9.7)$$

see Exercise 9.3. Hence, the delta always finishes at 1 for options that expire in-the-money and 0 for options that expire out-of-the-money. If $S(t) \approx E$ for times close to expiry, then the delta is liable to swing wildly between values at ≈ 1 (when $S(t)$ goes above E) and ≈ 0 (when $S(t)$ dips below E). Our next experiment illustrates this effect.

Computational example Here we repeat the computation that produced Figures 9.1 and 9.2 with the strike price reset to $E = 1.9$, so that the option frequently jumps in/out-of-the-money near expiry. Figure 9.3 shows that the corresponding delta value lurches dramatically as expiry is approached.   ◇

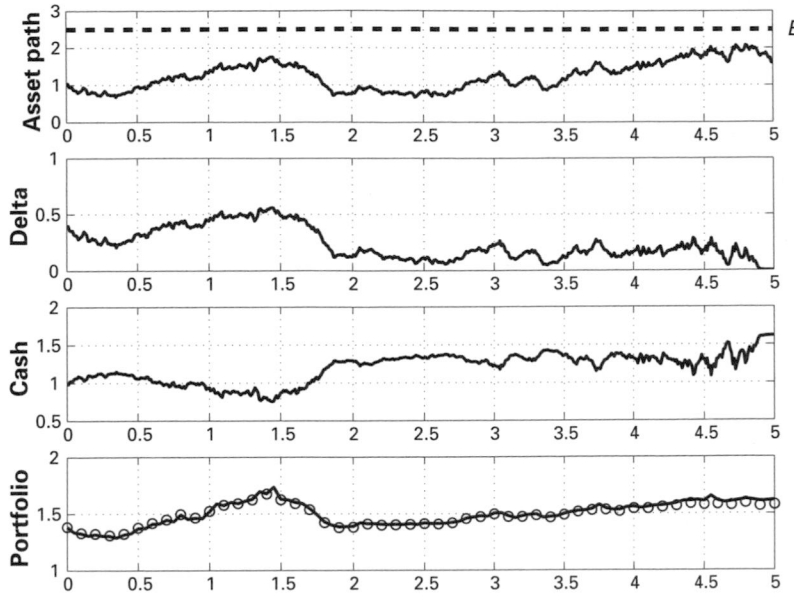

Fig. 9.2. Discrete hedging simulation. Option expires out-of-the-money. Upper: discrete asset path. Upper-middle: delta values (also asset holding in portfolio). Lower-middle: cash holding in portfolio. Lower: portfolio value (solid), theoretical portfolio value (9.5) (circles).

The delta behaviour near expiry that was observed in Figures 9.1 to 9.3, and is encapsulated in (9.7), has a simple financial interpretation. For $t \approx T$ there is little time left for the asset value to change – if it is currently in/out-of-the-money then it will probably remain in/out-of-the-money. In particular, if the call option is in-the-money then any upward or downward movement in the asset corresponds almost directly to the same upward or downward movement in the payoff. In other words, the call option and the asset are very highly correlated – they share the same risk. Since the portfolio is designed to replicate the risk in the option, it follows that it will hold approximately 1 unit of asset, so $\Delta_i \approx 1$. Conversely, if the call option is out-of-the-money close to expiry then the payoff is very likely to be zero whatever happens to the asset – there is no risk, so we should not be holding any asset.

The analogous results to (9.7) for a European put option are

$$\lim_{t \to T^-} \frac{\partial P(S, t)}{\partial S} = \begin{cases} 0, & \text{if } S(T) > E, \\ -\frac{1}{2}, & \text{if } S(T) = E, \\ -1, & \text{if } S(T) < E, \end{cases} \qquad (9.8)$$

see Exercise 9.4, and a similar financial argument applies, see Exercise 9.5.

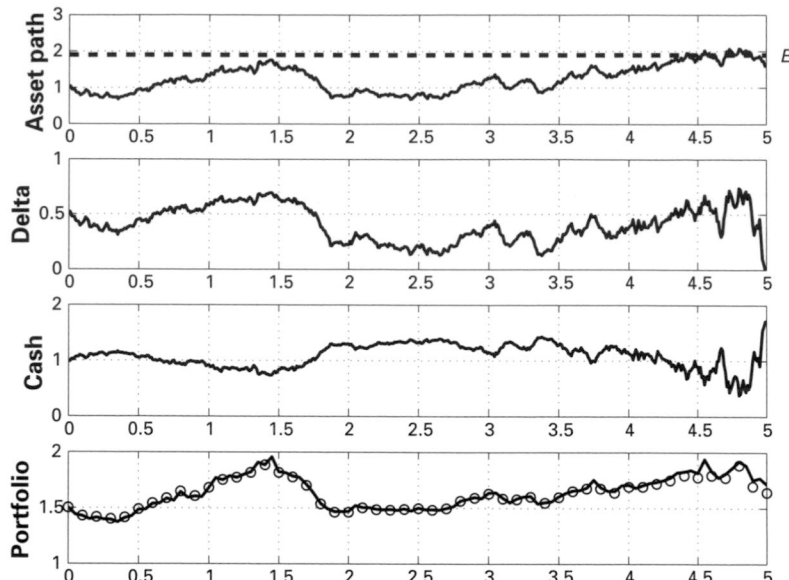

Fig. 9.3.   Discrete hedging simulation. Option expires almost at-the-money. Upper: discrete asset path. Upper-middle: delta values (also asset holding in portfolio). Lower-middle: cash holding in portfolio. Lower: portfolio value (solid), theoretical portfolio value (9.5) (circles).

9.4 Large-scale test

We finish with an experiment that looks at the success of discrete hedging over a large number of sample paths, and also illustrates that the option value is independent of the drift parameter, μ, in the asset price model.

Computational example   Here we take a European put option with $S_0 = 5$, $E = 5$, $r = 0.05$ and $\sigma = 0.3$, with $T = 3$. We computed 500 discrete asset paths with time-spacings $\delta t = 10^{-2}$. The upper picture in Figure 9.4 plots $S(T)$ on the horizontal axis against

$$\Pi(S(T), T) + (P(S_0, 0) - \Pi(S_0, 0)) \, e^{rT} \qquad (9.9)$$

on the vertical axis for the case $\mu = 0.02$. There are 500 such points, one for each asset path. We computed $P(S_0, 0)$ in (9.9) from the Black–Scholes formula (8.24). If the discrete hedging is successful, then an analogous identity to (9.5) holds for $P(S(t), t)$. In particular, it holds at expiry, so (9.9) should agree with the put payoff $\max(E - S(T), 0)$. This 'hockey stick' payoff curve is superimposed as a dashed line. We see that the dots lie close to the dashed line, and hence the discrete hedging algorithm behaves as predicted. The lower picture in

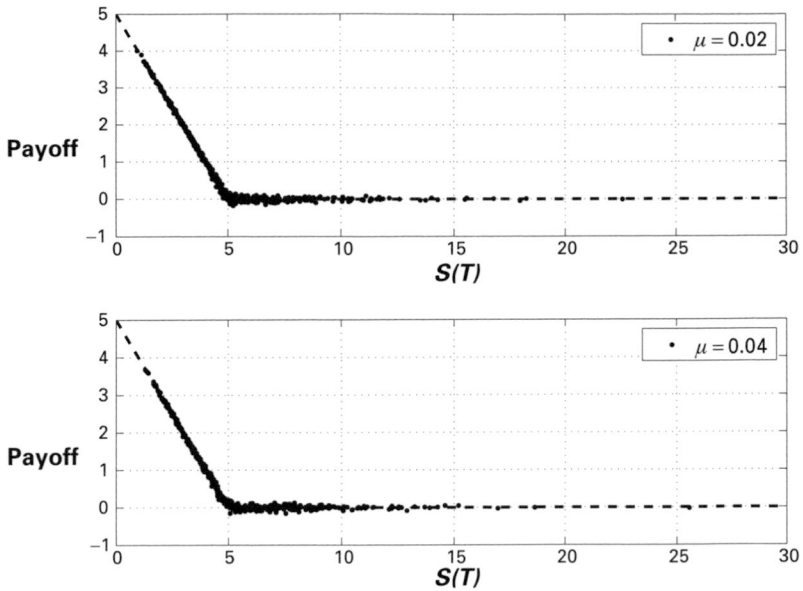

Fig. 9.4. Large-scale discrete hedging example for a European put. Dots represent normalized final payoff (9.9) for 500 asset paths. Exact hockey stick payoff is superimposed as a dashed line. Upper picture, $\mu = 0.02$. Lower picture, $\mu = 0.04$.

Figure 9.4 shows the same computations with μ changed to 0.04. This illustrates the phenomenon that the option value does not depend upon μ. ◇

9.5 Long-Term Capital Management

There are many instances of academics with an expertise in mathematical finance turning their hands to real-life trading. The most high-profile and, ultimately, sobering example involves Long-Term Capital Management (LTCM). This was a hedge fund that invested money supplied by its partners and a limited number of wealthy clients. Two of the partners, closely involved in day-to-day trading strategies, were Robert Merton and Myron Scholes – founding fathers of the 'rocket science' of option valuation theory. The fund, set up in 1994, was extremely successful at raising capital and for a period of around four years produced impressively high returns. Although sometimes referred to as an arbitrage unit, LTCM typically scoured the international markets looking for *low risk* opportunities to make relatively small percentage gains. The fund used *leverage* – investing borrowed money – to scale up these tiny margins into large profits. One commentator likened their trades to 'picking up nickels in front of bulldozers' (Lowenstein, 2001, page 102). At the peak of the fund's success, Merton and Scholes received their Nobel Prizes. However, in mid-1998 a combination of extreme events in the market plunged LTCM into deep trouble. One of the key difficulties they then faced was *illiquidity*. LTCM became desperate to offload a vast range of complicated portfolios, but the small set of potential buyers were, quite reasonably, holding out in the expectation that prices would drop further. (The assumption of liquidity – there always being a ready supply of buyers and sellers – is implicit in the Black–Scholes theory.) The bulldozers were moving in. The decline of LTCM and the enormity of its potential debts were brought to the attention of The Federal Reserve Bank of New York (the Fed), a major component of the US Federal Reserve System. Quite remarkably, the Fed became concerned that bankruptcy of LTCM could create such a hole that the overall stability of the market was at threat. Very rapidly, the Fed managed to persuade a consortium of major banks and investment houses to bail out LTCM in order to prevent the very real possibility of a total meltdown of the financial system.[1]

Overall, a dollar invested in LTCM grew to a height of around \$2.85, but dropped sharply to a paltry 23 cents, and the partners lost personal fortunes. A fast-paced and highly informative account of the LTCM debacle, with input from a number of first-hand witnesses, is given in (Lowenstein, 2001).

9.6 Notes

If you understand the hedging idea, it is perfectly reasonable for you to ask why options exist, that is,

given that it is possible to reproduce the payoff of an option using only cash and the underlying asset, why is there a market for options?

One answer is that the Black–Scholes theory relies on assumptions that are not universally valid, and it is neither convenient nor feasible for most of us to carry out hedging. On one side there is a large group of investors who view options as an excellent means to alleviate their exposure to risk, and another large group who see options as a great way to speculate on the market. On the other side there is a complementary group of well-connected players, with the resources to manipulate complicated portfolios and negotiate relatively small transaction costs, who are willing to accept the Black–Scholes value plus a small premium.

EXERCISES

9.1.   ⋆ Show from (9.1) and (9.2) that $\partial C/\partial S > 0$ and $\partial P/\partial S < 0$.

---

[1] Lowenstein (Lowenstein, 2001, page 198) quotes Sandy Warner from J. P. Morgan: 'Boys, we're going to a picnic and the tickets cost \$250 million'.

```
%CH09      Program for Chapter 9
%
% Illustrates delta hedging by computing an approximate
% replicating portfolio for a European call
%
% Portfolio is 'asset' units of asset and an amount 'cash' of cash
% Plot actual and theoretical portfolio values

randn('state',100)
clf

%%%%%%%%% Problem parameters %%%%%%%%%%%%%%
Szero = 1; sigma = 0.35; r = 0.03; mu = 0.02; T = 5; E = 2;
Dt = 1e-2; N = T/Dt; t = [0:Dt:T];
%%%%%%%%%%%%%%%%%%%%%%%%%%%%%%%%%%%%%%%%%%

S = zeros(N,1); asset = zeros(N,1); cash = zeros(N,1);
portfolio = zeros(N,1); Value = zeros(N,1);

[C,Cdelta,P,Pdelta] = ch08(Szero,E,r,sigma,T-t(1));

S(1) = Szero;
asset(1) = Cdelta;
Value(1) = C;
cash(1) = 1;
portfolio(1) = asset(1)*S(1) + cash(1);

for i = 1:N
   S(i+1) = S(i)*exp((mu-0.5*sigma^2)*Dt+sigma*sqrt(Dt)*randn);
   portfolio(i+1) = asset(i)*S(i+1) + cash(i)*(1+r*Dt);
   [C,Cdelta,P,Pdelta] = ch08(S(i+1),E,r,sigma,T-t(i+1));
   asset(i+1) = Cdelta;
   cash(i+1) = cash(i)*(1+r*Dt) - S(i+1)*(asset(i+1) - asset(i));
   Value(i+1) = C;
end

Vplot = Value - (Value(1) - portfolio(1))*exp(r*t)';
plot(t(1:5:end),Vplot(1:5:end),'bo')
hold on
plot(t(1:5:end),portfolio(1:5:end),'r-','LineWidth',2)
xlabel('Time'), ylabel('Portfolio')
legend('Theoretical Value','Actual Value')
grid on
```

Fig. 9.5.   Program of Chapter 9: ch09.m.

9.2. ★★ By making reference to the limit definition

$$\frac{\partial C}{\partial S} = \lim_{\delta S \to 0} \left(\frac{C(S + \delta S, t) - C(S, t)}{\delta S} \right),$$

give an intuitive reason why $\partial C/\partial S \geq 0$. Do the same for $\partial P/\partial S \leq 0$.

9.3. ★★ Using the expression (9.1), confirm the limiting behaviour for $\partial C(S, t)/\partial S$ displayed in (9.7).

9.4. ★★ Using the expression (9.2), confirm the limiting behaviour for $\partial P(S, t)/\partial S$ displayed in (9.8).

9.5. ★★ Give a financial argument that explains why $\partial P(S, t)/\partial S \to -1$ at expiry for an in-the-money put option and $\partial P(S, t)/\partial S \to 0$ at expiry for an out-of-the-money put option.

9.7 Program of Chapter 9 and walkthrough

Our program ch09 implements a discrete hedging simulation and produces a picture like the lower plots in Figures 9.1–9.3. It is listed in Figure 9.5. Here, S, asset, Value and cash are N by 1 arrays whose ith entries store the asset price, asset holding, Black–Scholes option value and cash holding at time t(i), respectively. After initializing parameters, we set up a for loop that updates the portfolio as described in Section 9.2. The Black–Scholes function ch08 from Chapter 8 is used to find the option value and the delta.

On exiting the loop we superimpose the left- and right-hand sides of (9.5), plotting at every fifth time point.

PROGRAMMING EXERCISES

P9.1. Adapt ch09.m to investigate how the average discrepancy at expiry, (9.6), varies as a function of δt.

P9.2. Perform a large-scale test for a call option in the style of Figure 9.4.

Quotes

The professors were brilliant at reducing a trade to pluses and minuses;
they could strip a ham sandwich to its component risks;
but they could barely carry on a normal conversation.

ROGER LOWENSTEIN (Lowenstein, 2001)

After closing about 200 000 option transactions
(that is separate option tickets)
over 12 years and studying about 70 000 risk management reports,
I felt that I needed to sit down and reflect on the thousands
of mishedges I had committed.

NASSIM TALEB (Taleb, 1977)

It is probably safe to say that
the derivatives industry would be stuck in the psychedelic 60s,
and many talented mathematicians would still be
teaching freshman algebra for $20,000 a year
had Black, Scholes, and Merton not made their contribution.

DON M. CHANCE, 'Rethinking Implied Volatility' *Financial Engineering News*, January/February 2003.

10

The Greeks

10.1 Motivation

The Black–Scholes option valuation formulas (8.19) and (8.24) depend upon S, t and the parameters E, r and σ. In this chapter we derive expressions for partial derivatives of the option values with respect to these quantities. These results are useful for a number of reasons.

- Traders like to know the sensitivity of the option value to changes in these quantities. The sensitivities can be measured by these partial derivatives; see Exercise 10.1.
- Computing the partial derivatives allows us to confirm that the Black–Scholes PDE has been solved.
- Examining the signs of the derivatives gives insights into the underlying formulas.
- The derivative $\partial V / \partial S$ is needed in the delta hedging process.
- The derivative $\partial V / \partial \sigma$ comes into play in Chapter 14, where we compute the implied volatility.

We focus on the case of a call option. Exercise 10.7 asks you to do the same things for a put.

10.2 The Greeks

Certain partial derivatives of the option value are so widely used that they have been assigned Greek names and symbols,[1]

$$\Delta := \frac{\partial C}{\partial S} \qquad \text{(delta)},$$

[1] Vega is not actually a Greek name, and does not even qualify for a symbol.

$$\Gamma := \frac{\partial^2 C}{\partial S^2} \quad \text{(gamma)},$$

$$\rho := \frac{\partial C}{\partial r} \quad \text{(rho)},$$

$$\Theta := \frac{\partial C}{\partial t} \quad \text{(theta)},$$

$$\text{vega} := \frac{\partial C}{\partial \sigma} \quad \text{(vega)}.$$

By differentiating C in (8.19), using (8.20) and (8.21), it is possible to find explicit expressions for these quantities. Before launching into this process we make note of two useful facts. First, it follows from (3.18) that

$$N'(x) = \frac{1}{\sqrt{2\pi}} e^{-\frac{1}{2}x^2}.$$

Our second fact,

$$SN'(d_1) - e^{-r(T-t)} E N'(d_2) = 0, \tag{10.1}$$

is to be proved in Exercise 10.2.

Differentiating with respect to S in (8.19) we have

$$\Delta = N(d_1) + SN'(d_1)\frac{\partial d_1}{\partial S} - Ee^{-r(T-t)}N'(d_2)\frac{\partial d_2}{\partial S}$$

$$= N(d_1) + \frac{N'(d_1)}{\sigma\sqrt{T-t}} - Ee^{-r(T-t)}\frac{N'(d_2)}{S\sigma\sqrt{T-t}}.$$

Appealing to (10.1), we see that the second and third terms on the right-hand side cancel, giving

$$\Delta = N(d_1), \tag{10.2}$$

a result that we used in Chapter 9. We then have

$$\Gamma = \frac{\partial \Delta}{\partial S} = N'(d_1)\frac{\partial d_1}{\partial S} = \frac{N'(d_1)}{S\sigma\sqrt{T-t}}. \tag{10.3}$$

Next, differentiating C with respect to r we find that

$$\rho := \frac{\partial C}{\partial r} = SN'(d_1)\frac{\partial d_1}{\partial r} + (T-t)Ee^{-r(T-t)}N(d_2) - Ee^{-r(T-t)}N'(d_2)\frac{\partial d_2}{\partial r}$$

$$= SN'(d_1)\frac{T-t}{\sigma\sqrt{T-t}} + (T-t)Ee^{-r(T-t)}N(d_2)$$

$$- Ee^{-r(T-t)}N'(d_2)\frac{T-t}{\sigma\sqrt{T-t}}.$$

As before, (10.1) allows us to cancel terms, and we find that

$$\rho = (T - t)Ee^{-r(T-t)}N(d_2).$$ (10.4)

Similar analysis shows that

$$\Theta = \frac{-S\sigma}{2\sqrt{T - t}}N'(d_1) - rEe^{-r(T-t)}N(d_2)$$ (10.5)

and

$$\text{vega} = S\sqrt{T - t}N'(d_1),$$ (10.6)

see Exercises 10.3 and 10.4.

10.3 Interpreting the Greeks

It is possible to interpret some of the Greek formulas from a financial viewpoint and to check that they agree with intuition.

First we recall that the limiting behaviour of delta was characterized and interpreted in Section 9.3. We also know from Exercise 9.1 that $\Delta > 0$ up to expiry. This makes sense, because an increase in the asset price increases the likely profit at expiry.

From (10.4) we see that $\rho > 0$ before expiry. To explain this we note that increasing the interest rate is equivalent to lowering the exercise price E. (The value of a fixed amount E at some fixed time in the future becomes less if the interest rate increases.) This makes a payoff more likely, which increases the value of the option.

The expression (10.5) shows that $\Theta < 0$. This property could also be deduced directly from the general, asset-model-independent argument in Section 2.6 concerning the monotonicity of the time-zero call option value with respect to the expiry date, see Exercise 10.5.

The vega in (10.6) is always positive before expiry. This can be understood by considering that an increase in volatility leads to a wider spread of asset prices. However, assets moving deeper out-of-the-money have no effect on the option price (the payoff remains zero) while assets moving deeper into-the-money lead to a greater payoff. Because of this asymmetry, increasing σ has a net positive effect. We return to vega in Chapter 14.

10.4 Black–Scholes PDE solution

Having worked out the partial derivatives, we are in a position to confirm that $C(S, t)$ in (8.19) satisfies the Black–Scholes PDE (8.15). Using our expressions for Δ, Γ, ρ and Θ, we have

$$\frac{\partial C}{\partial t} + \frac{1}{2}\sigma^2 S^2 \frac{\partial^2 C}{\partial S^2} + rS\frac{\partial C}{\partial S} - rC = \frac{-S\sigma}{2\sqrt{T-t}}N'(d_1) - rEe^{-r(T-t)}N(d_2)$$

$$+ \frac{1}{2}\sigma^2 S^2 \frac{N'(d_1)}{S\sigma\sqrt{T-t}} + rSN(d_1)$$

$$- r\left(SN(d_1) - Ee^{-r(T-t)}N(d_2)\right)$$

$$= 0.$$

10.5 Notes and references

Many texts present the formulas for the Greeks without getting into the nitty-gritty of differentiation. Exceptions are (Kwok, 1998; Nielsen, 1999). For more information on interpreting the Greek formulas, see (Hull, 2000; Kwok, 1998; Nielsen, 1999), for example.

EXERCISES

10.1. ★ If $F : \mathbb{R} \to \mathbb{R}$ is differentiable, use the definition of the differentiation process to explain why $F'(x)$ measures the sensitivity of F to changes in x.

10.2. ★★ Verify the identity

$$\log\left(\frac{SN'(d_1)}{e^{-r(T-t)}EN'(d_2)}\right) = 0,$$

and hence derive (10.1).

10.3. ★★★ Establish (10.5) and (10.6).

10.4. ★★ Give a financial explanation why $\Delta < 0$ for a put option (proved in Exercise 9.1).

10.5. ★★ Show that the condition $\partial C/\partial t \le 0$ can be deduced directly from the conclusion in Section 2.6 that the time-zero call option value is a non-decreasing function of the expiry date.

10.6. ★★★ Using (10.1), show that the partial derivative $\partial C/\partial E$ (which, sadly, does not have a Greek name) satisfies

$$\frac{\partial C}{\partial E} = -e^{-r(T-t)}N(d_2).$$

Deduce that $\partial C/\partial E < 0$ and interpret this result.

10.7. ★★★ Using the put–call parity identity (8.23), for each expression for a partial derivative of C that appears in this chapter obtain an expression

```
function [C, Cdelta, Cvega, P, Pdelta, Pvega] = ch10(S,E,r,sigma,tau)
% Program for Chapter 10
% This is a MATLAB function
%
% Input arguments: S = asset price at time t
%                  E = exercise price
%                  r = interest rate
%                  sigma = volatility
%                  tau = time to expiry (T-t)
%
% Output arguments: C = call value, Cdelta = delta value of call
%                   Cvega = vega value of call
%                   P = Put value, Pdelta = delta value of put
%                   Pvega = vega value of put
%
%   function [C, Cdelta, Cvega, P, Pdelta, Pvega] = ch10(S,E,r,sigma,tau)

if tau > 0
  d1 = (log(S/E) + (r + 0.5*sigma^2)*(tau))/(sigma*sqrt(tau));
  d2 = d1 - sigma*sqrt(tau);
  N1 = 0.5*(1+erf(d1/sqrt(2)));
  N2 = 0.5*(1+erf(d2/sqrt(2)));
  C = S*N1-E*exp(-r*(tau))*N2;
  Cdelta = N1;
  Cvega = S*sqrt(tau)*exp(-0.5*d1^2)/sqrt(2*pi);
  P = C + E*exp(-r*tau) - S;
  Pdelta = Cdelta - 1;
  Pvega = Cvega;
else
  C = max(S-E,0);
  Cdelta = 0.5*(sign(S-E) + 1);
  Cvega = 0;
  P = max(E-S,0);
  Pdelta = Cdelta - 1;
  Pvega = 0;
end
```

Fig. 10.1.   Program of Chapter 10: ch10.m.

for the corresponding partial derivative of P. For each discussion of the sign of a partial derivative of the call option value, give a discussion of the corresponding sign for the put. In particular, show by example that $\partial P/\partial t$ may be positive or negative. Using your expressions, confirm that $P(S, t)$ satisfies the Black–Scholes PDE (8.15).

10.6 Program of Chapter 10 and walkthrough

The program ch10, listed in Figure 10.1, is an extended version of the function ch08 that returns values of the call and put vega. These values will be needed by the program ch14 in Chapter 14. The call vega formula is given by (10.6) and the put vega formula was derived in Exercise 10.7.

An example of the function in use is

```
>> S = 2; E = 2.5; r = 0.03; sigma = 0.25; tau = 1;
>> [C, Cdelta, Cvega, P, Pdelta, Pvega] = ch10(S,E,r,sigma,tau)
```

which outputs

```
C = 0.0691
Cdelta = 0.2586
Cvega = 0.6470
P = 0.4953
Pdelta = -0.7414
Pvega = 0.6470
```

PROGRAMMING EXERCISES

P10.1. Adapt function ch10.m to return more Greeks.

P10.2. Investigate the use of MATLAB's symbolic toolbox to confirm the results in this chapter.

Quotes

Proof: Use the Black–Scholes formula (6.46) and take derivatives.
The (brave) reader is invited to carry this out in detail.
The calculations are sometimes quite messy.
                         THOMAS BJÖRK (on calculating the Greeks) (Björk, 1998)

I am so glad I am a Beta,
the Alphas work so hard.
And we are much better than the Gammas and Deltas.
                         ALDOUS HUXLEY from *Brave New World*, 1932 (1894–1963)

You can overintellectualize these Greek letters.
One Greek word that ought to be in there is *hubris*.
                         DAVID PFLUG, source (Lowenstein, 2001)

Neither Black nor Scholes,
at first,
knew how to derive the solution to these complicated equations, ...
           MARK. P. KRITZMAN (Kritzman, 2000) with reference to the Black–Scholes PDE

11

More on the Black–Scholes formulas

11.1 Motivation

We now take the opportunity to reflect a little more on the Black–Scholes option valuation formulas. In particular, Figure 11.3 is an attempt to squeeze everything we have learnt into a single picture.

11.2 Where is μ?

The Black–Scholes formulas allow us to determine a fair price at time zero for a European call or put option in terms of the initial asset price, S_0, the exercise price, E, the asset volatility, σ, the risk-free interest rate, r, and the expiry date, T. Each of these quantities is known, with the exception of the asset volatility, σ. Chapters 14 and 20 are concerned with the task of estimating σ using information available from the market. A big surprise, and perhaps the most remarkable aspect of the Black–Scholes theory, is that the option price does not depend on the drift parameter, μ, which, from (6.11), determines the expected growth of the asset. A consequence is that two investors could have wildly different views about what is an appropriate value of μ for a particular asset and yet, if they agreed on the volatility and accepted the assumptions that go into the Black–Scholes analysis, they would come up with the same value for the option. This phenomenon, which may seem highly questionable at first glance, is a consequence of the fact that Black–Scholes determines a **fair** value for the option – a value that can be recovered using the risk-free delta hedging strategy and hence the value, in the presence of arbitrageurs, that the forces of supply and demand dictate for the market.

Suppose that there are two speculators,

- Speculator A, who believes that the asset price will follow (6.9) with drift μ_A and volatility σ, and
- Speculator B, who believes that the asset price will follow (6.9) with drift μ_B and volatility σ.

Suppose the speculators wish to take a naked, long position on a European call option – that is, they wish to buy the option without performing any accompanying hedging. If $\mu_A \gg \mu_B$ then, presumably, Speculator A would find the Black–Scholes option value more attractive than Speculator B. This does not contradict the previous theory. A speculator who is *willing to accept some risk* may value an option differently to the Black–Scholes formula. However, if you are selling the option and wish to hedge in order to *eliminate risk* (and if you believe in the Black–Scholes assumptions) then (8.19) and (8.24) are the relevant values.

11.3 Time dependency

Figure 11.1 shows the Black–Scholes values of a call and a put option, as functions of asset price S, for certain fixed times t. We used $E = 1, r = 0.05, \sigma = 0.6$ and took expiry date $T = 1$. Figure 11.2 shows the same information in three-dimensional form.

In both cases, we see that as t approaches the expiry date T, the option value approaches the hockey-stick payoff function. This will always be the case, as we showed in Exercise 8.1. In the case of a call option, for each S, the value appears to converge to the hockey stick monotonically from above as t approaches expiry. This is also generic, since, as we saw in Section 10.3, the time derivative, theta, is always negative. On the other hand, for the put option, the convergence is not uniformly from above or below. This is consistent with Exercise 10.7, where you were asked to show that a put's time derivative can be negative or positive, see Exercise 11.2.

11.4 The big picture

Figure 11.3 draws the Black–Scholes European call option value, $C(S, t)$, as a surface above the (S, t)-plane, This emphasizes that $C(S, t)$ is a **smooth** function of S and t. Onto the $C(S, t)$-surface a solid white line adds the corresponding $C(S_i, t_i)$ values mapped out by a discrete asset path. This picture illustrates that

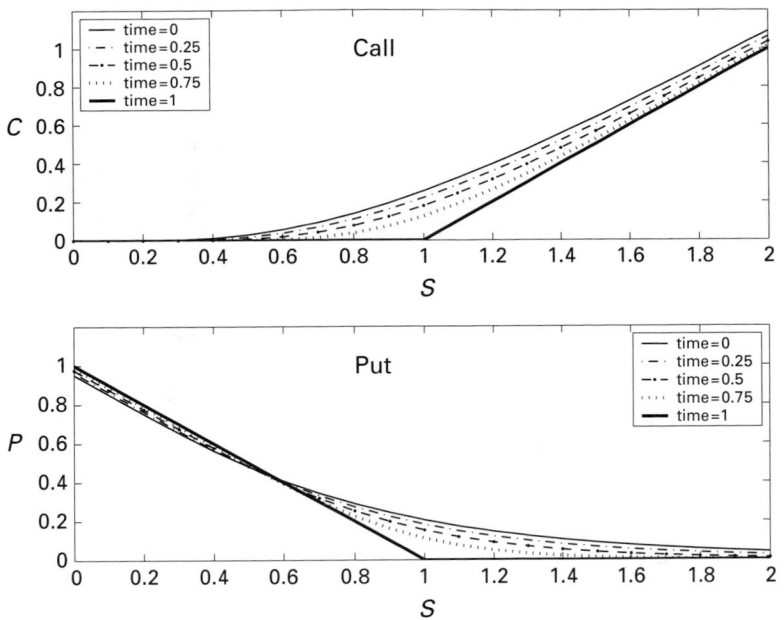

Fig. 11.1. Option value in terms of asset price at five different times. Upper: European call. Lower: European put.

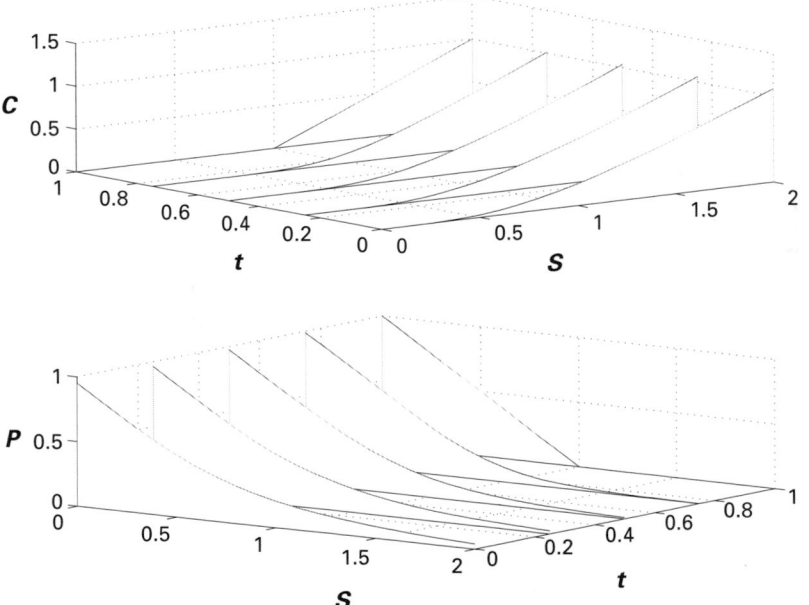

Fig. 11.2. Three-dimensional version of Figure 11.1.

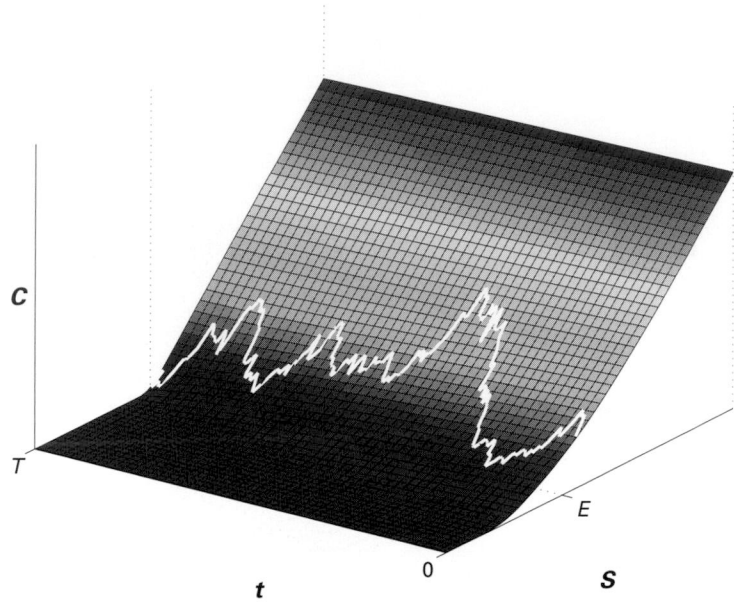

Fig. 11.3.   European call: Black–Scholes surface with asset path superimposed.

- the Black–Scholes option value surface is smooth,
- an asset path is jagged,
- as time varies, an asset path maps out a jagged 'option path' over the smooth option value surface.

Figure 11.4 repeats the exercise for a put option.

In Figure 11.5 we plot the delta surface, $\partial C / \partial S$, for a call option and superimpose three option paths. One option expires in-, one out-of- and one almost at-the-money. As discussed in Section 9.3, the rapid gradient of the delta surface induces large variations (and hence large swings in the amount of asset in the replicating portfolio) when the option is close to being at-the-money. Note from (9.1)–(9.2) that, since the vertical axis in the figure has no markings, the corresponding picture for a put option would be identical.

11.5 Change of variables

On the face of it, the Black–Scholes value of a European call or put option depends on the strike price, E, the expiry time, T, the volatility σ and the interest rate, r, as well as the asset price S and time t. However, by a judicious re-scaling, we can reduce the length of this list to two.

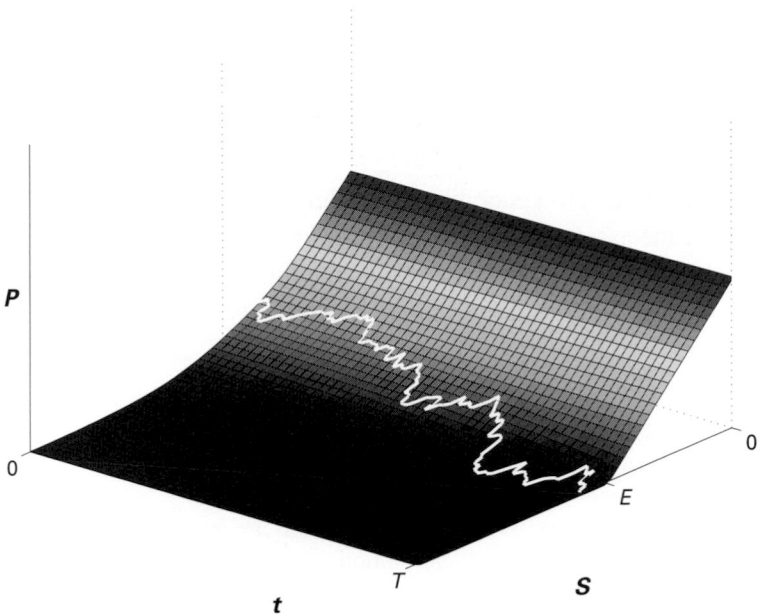

Fig. 11.4. European put: Black–Scholes surface with asset path superimposed.

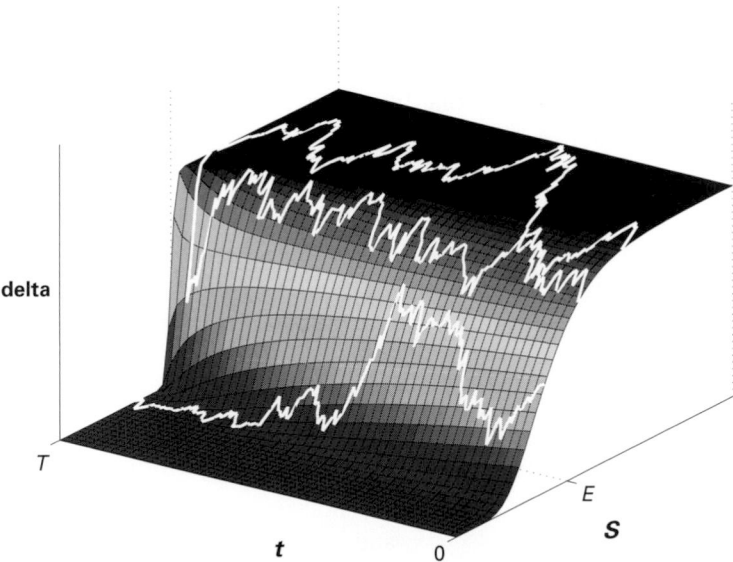

Fig. 11.5. Black–Scholes surface for delta with three asset paths superimposed.

We will introduce three new dimensionless quantities. First is the *moneyness ratio*

$$m := \log \frac{Se^{r(T-t)}}{E}.$$

To interpret m, we need to generalize (6.11) into the formula $Se^{\mu(T-t)}$ for the expected value of the asset at expiry, given asset price S at time t, Now we make the assumption that the asset growth rate equals the interest rate, $\mu = r$. This assumption will be examined in detail in Chapter 12; for now, we simply note that it leads to the following conclusions.

> If $m > 0$, then the expected asset value at expiry is greater than the strike price. In a 'risk-neutral expectation at expiry' sense, a call option is in-the-money and a put option is out-of-the-money.
> If $m = 0$, then, in the same sense, call and put options are at-the-money.
> If $m < 0$, then, in the same sense, a call option is out-of-the-money and a put option is in-the-money.

Second, we have the *scaled volatility*

$$\widehat{\tau} := \sigma\sqrt{T-t}.$$

Here, the volatility is combined with the square root of the time to expiry. This is natural, since, for example, volatility appears in the form $\sigma^2(t_{i+1}-t_i)$ in the underlying asset model (6.9). The third step is to scale the option values by the asset price, by letting

$$c := \frac{C}{S}, \qquad \text{for a call option,}$$

and

$$p := \frac{P}{S}, \qquad \text{for a put option.}$$

In these new variables, d_1 and d_2 in (8.20) and (8.21) simplify to

$$d_1 = \frac{m}{\widehat{\tau}} + \frac{\widehat{\tau}}{2} \qquad \text{and} \qquad d_2 = \frac{m}{\widehat{\tau}} - \frac{\widehat{\tau}}{2}, \tag{11.1}$$

and, from (8.19) and (8.24), the re-scaled call and put values become

$$c(m,\widehat{\tau}) = N(d_1) - e^{-m}N(d_2) \qquad \text{and} \qquad p(m,\widehat{\tau}) = e^{-m}N(-d_2) - N(-d_1), \tag{11.2}$$

see Exercise 11.3.

11.6 Notes and references

Colour versions of Figures 11.3, 11.4 and 11.5 can be downloaded from this book's website, mentioned in the preface.

EXERCISES

11.1. ⋆⋆ Consider the following 'explanation' of why the Black–Scholes European call option value curve $C(S, t)$ lies above the payoff hockey stick $\max(S(t) - E, 0)$, for $t < T$.

Since $\mathbb{E}(S(t)) = S_0 e^{\mu t}$, the asset price generically drifts upwards. Hence, on average, the asset price will increase between time t and expiry, so the time t value is greater than $\max(S(t) - E, 0)$.

Is this argument valid?

11.2. ⋆⋆ Show how Exercise 10.7 provides a counterexample to the following statement:

As t goes from 0 to T, the Black–Scholes European put option value always approaches the payoff hockey-stick function from below.

11.3. ⋆⋆ Verify (11.1) and (11.2).

11.4. ⋆ ⋆ ⋆ In the case where the volatility, σ, is zero in the asset model (6.9), the final asset price is the nonrandom quantity $S_0 e^{\mu T}$. The payoff from a European option is then guaranteed to be $\max(S_0 e^{\mu T} - E, 0)$. It may thus be argued that the time-zero option value **must** be $e^{-rT} \max(S_0 e^{\mu T} - E, 0)$. However, this value clearly depends upon μ, whilst the Black–Scholes formula does not. (In fact, looking ahead to (14.2), the Black–Scholes value is $e^{-rT} \max(S_0 e^{rT} - E, 0)$.) Can you resolve this apparent contradiction?

11.5. ⋆⋆ Show that 'Call$(-\sigma) = -$Put(σ)', that is, replacing σ in (8.19) by $-\sigma$ is equivalent to evaluating $-P(S, t)$ in (8.24). This relation is sometimes called *put–call supersymmetry*.

11.7 Program of Chapter 11 and walkthrough

The program ch11 plots the Black–Scholes surface above the (S, t)-plane for a European call, in the style of Figure 11.3. It is listed in Figure 11.6. We initialize E,r,sigma and T, and set up the array Svals of 50 equally spaced asset prices between 0 and 3 and the array tvals of 50 equally spaced time points between 0 and T. The nested for loops then work through Svals and tvals, using ch08 to evaluate the Black–Scholes formula. The European call value is stored in the two-dimensional array Call. We then use meshgrid to set up two-dimensional arrays Smat and tmat that are appropriate for use with the three-dimensional plotting function mesh.

```
%CH11     Program for Chapter 11
%
% Draws Black-Scholes surface for European call

clf

%%%%%%% Problem parameters %%%%%%%%%
E = 1;    r = 0.05;    sigma = 0.2;   T = 1;    L =50;
%%%%%%%%%%%%%%%%%%%%%%%%%%%%%%%%%%

Svals = linspace(0,3,L);
tvals = linspace(0,T,L);
C = zeros(L,L);
for i = 1:L
   S = Svals(i);
   for j = 1:L
     t = tvals(j);
     [Call,Calldelta,Put,Putdelta] = ch08(S,E,r,sigma,T-t);
     C(i,j) = Call;
   end
end

[Smat,tmat] = meshgrid(Svals,tvals);
mesh(Smat,tmat,C')
xlabel('S'), ylabel('t'), zlabel('C(S,t)')
```

Fig. 11.6.   Program of Chapter 11: ch11.m.

PROGRAMMING EXERCISES

P11.1. Edit ch11.m so that it applies to a European put option, as in Figure 11.4.

P11.2. Edit ch11.m so that it applies to the delta of a European call option, as in Figure 11.5, and investigate the use of surf, surfc and waterfall instead of mesh.

Quotes

The Black–Scholes formula is still around,
even though it depends on at least 10 unrealistic assumptions.
Making the assumptions more realistic
hasn't produced a formula that works better across a wide range of circumstances.

FISCHER BLACK (Black, 1989)

We know this doesn't work by rote.
But this is the best model we have.
You look at the old-timers who went with their gut.
You had this model, you had these numbers,

and in the end you thought they were a lot more powerful than a guy's gut.

ROBERT STAVIS, former member of the Arbitrage group at Salomon Brothers, source
(Lowenstein, 2001)

A first-rate theory predicts,
a second-rate theory forbids
and a third-rate theory explains after the event.

ALEXANDER KITAIGORODSKI, 1975,
source www.byrneweb.com/sunburn/quotes. html

12

Risk neutrality

12.1 Motivation

In the days before the Black–Scholes formula, it was often argued that a reasonable way to value an option is to take the *expected payoff*. In this chapter we show how the expected payoff idea fits in with the Black–Scholes methodology. This leads us to the concept of *risk neutrality*, which will play a fundamental role in Chapters 15, 16 and beyond, when we discuss computational algorithms.

12.2 Expected payoff

To cover European call and put options in a single notation, we let $\Lambda(x)$ denote the payoff function, so $\Lambda(x) = \max(x - E, 0)$ for a call and $\Lambda(x) = \max(E - x, 0)$ for a put. The treatment here easily generalizes to other *European-style options*, that is, options whose payoff may be expressed as a function of the asset price at expiry.

Under our model (6.8), the final asset price, $S(T)$, is a random variable of the form $S(T) = S_0 e^{(\mu - \sigma^2/2)T + \sigma \sqrt{T} Z}$, where $Z \sim N(0, 1)$. So the payoff, $\Lambda(S(T))$, is also a known random variable. Why don't we simply take the time-zero option value to be the average payoff, suitably discounted for interest? This gives a value

$$e^{-rT} \mathbb{E}(\Lambda(S(T))). \tag{12.1}$$

Using (3.8) and the density function (6.10), this may be written

$$e^{-rT} \int_0^\infty \frac{\Lambda(x)}{x\sigma\sqrt{2\pi}\sqrt{T}} \exp\left(-\frac{\left(\log x - \log S_0 - (\mu - \frac{1}{2}\sigma^2)T\right)^2}{2\sigma^2 T}\right) dx. \tag{12.2}$$

115

More generally, we could regard the option value at asset price S and time t as the, suitably discounted, expectation of the payoff. Letting $W(S, t)$ denote this value, we have

$$W(S, t) = e^{-r(T-t)} \mathbb{E}\left(\Lambda(S(T)), \text{ given asset price } S \text{ at time } t\right), \qquad (12.3)$$

which may be written more explicitly as

$$W(S, t) = e^{-r(T-t)} \int_0^\infty \frac{\Lambda(x)}{x\sigma\sqrt{2\pi}\sqrt{T-t}}$$
$$\exp\left(-\frac{\left(\log x - \log S - (\mu - \frac{1}{2}\sigma^2)(T-t)\right)^2}{2\sigma^2(T-t)}\right) dx. \quad (12.4)$$

The values (12.2) and (12.4) are certainly relevant to an individual who is in the habit of writing or holding naked options. However, in comparison with the Black–Scholes approach to finding a *fair* option value, there are a number of related points to make.

(i) Formulas (12.2) and (12.4) were derived *without any reference to the idea of hedging to eliminate risk.*
(ii) Formulas (12.2) and (12.4) were derived *without any reference to the no arbitrage principle.*
(iii) Unlike the Black–Scholes PDE (8.15), the formulas (12.2) and (12.4) *depend on the parameter μ.*

Now the Black–Scholes theory tells us that there is only one fair value, and this must be the figure quoted in the market. If the market placed the option lower/higher, arbitrageurs would swoop en masse, buying/selling the option, delta hedging until expiry, and hence guaranteeing a riskless profit. The forces of supply and demand therefore constrain the option to the Black–Scholes level. It follows from point (iii) that the expected payoff approach cannot be used to get a fair value.

On the face of it, expected payoff seems to have no place in option valuation theory. However, by a remarkable twist, it is possible to rehabilitate the idea.

12.3 Risk neutrality

Figure 12.1 confirms that the time-zero discounted expected payoff (12.2) is indeed a function of μ. The solid line plots (12.2) as μ varies from 0 to 0.1 for a European call with $S_0 = 10$, $E = 9$, $r = 0.05$, $\sigma = 0.2$ and $T = 3$. As we would guess, the expected payoff increases with the growth rate, μ. Superimposed on the picture as a dashed line is the Black–Scholes option value, 2.66.

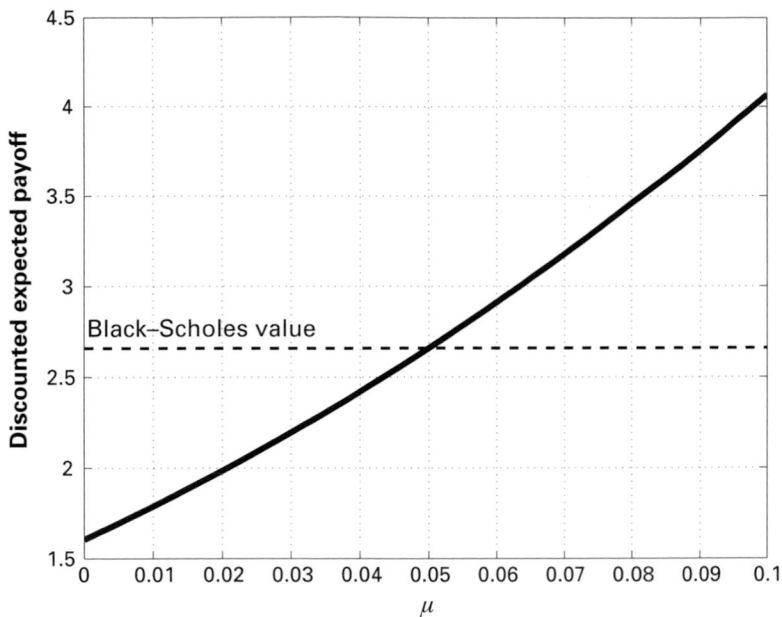

Fig. 12.1. Time-zero discounted expected payoff (12.2) for a European call. Black–Scholes value superimposed as a dashed line.

Keen-eyed observers will note that the solid curve in Figure 12.1 appears to pass through the Black–Scholes level at the value $\mu = r = 0.05$; that is, when the growth rate parameter matches the interest rate. This turns out to be no coincidence. Exercise 12.1 asks you to verify the general result that

$W(S, t)$ in (12.4) satisfies the Black–Scholes PDE (8.15) when $\mu = r$.

Now we check the final time and boundary conditions. Taking $t = T$ in (12.3), we note that if $S(T)$ is given, and thus nonrandom, then $\mathbb{E}(\Lambda(S(T))) = \Lambda(S(T))$, giving

$$W(S, T) = \Lambda(S(T)).$$

Hence the conditions (8.16) for a call and (8.25) for a put are satisfied. Similarly, if $S = 0$ at any time then we know from (6.9) that $S(T) = 0$, and hence in (12.3)

$$W(0, t) = e^{-r(T-t)}\Lambda(0).$$

This matches (8.17) and (8.26) for the call and put, respectively. Finally, we note that the arguments given to justify (8.18) and (8.27) are equally valid for (12.3). Overall, since $W(S, t)$ with $\mu = r$ satisfies the same PDE and the same final time/boundary conditions, the uniqueness of the solution tells us that

$W(S, t)$ in (12.4) reproduces the Black–Scholes option value when $\mu = r$.

We could re-write this conclusion as follows.

No matter what parameters μ and σ in the asset model (6.9) we believe to be correct, we can obtain the Black–Scholes option value by pretending that the drift, μ, is equal to the interest rate, r, and taking the discounted expected payoff.

In setting $\mu = r$ we are making what is known as a *risk neutrality* assumption.

We will see in Chapters 15 and 16 that the risk-neutral expectation framework allows us to develop computational methods for approximating options where analytical formulas are not available.

12.4 Notes and references

It is perfectly standard, but not particularly enlightening, to give the name *risk neutrality* to the condition $\mu = r$. The phrase borrows from the concept of a *risk-neutral investor*; an unlikely person who regards

- an investment with guaranteed rate of return r, and
- a risky investment with expected rate of return r

as equally attractive. In the case where all assets satisfy the lognormal model (6.9) with the same growth parameter μ – the so-called *risk-neutral world* – we see from (6.11) that a risk-neutral investor would have no preferences between investing in a bank and in any asset.

In the risk-neutral world, (6.11) shows that $\mathbb{E}(S(t)) = S_0 e^{rt}$, so the expected discounted asset price is $\mathbb{E}(e^{-rt} S(t)) = S_0$. In other words, the expected discounted asset price does not change with time; it remains at its time-zero level. A process like this, whose expected future value is given by its current value, is called a *martingale*. By using martingale theory it is possible to convert the simple observation in Exercise 12.1 into a rigorous and powerful theory for option valuation. In particular, this is an alternative way to derive the Black–Scholes formulas. The texts (Duffie, 2001; Karatzas and Shreve, 1998; Nielsen, 1999) cover this material in depth, while perhaps the most accessible introduction is (Baxter and Rennie, 1996). Chapter 6 of (Kritzman, 2000) also gives a very readable, example-driven coverage of risk neutrality.

In Chapter 16 we introduce the binomial method as a computational technique for option valuation. It is also possible to use the binomial framework as an analytical tool with which the Black–Scholes formulas can be derived without recourse to PDEs. The concept of risk neutrality arises quite naturally in this setting. Exercise 12.5 provides a cut-down version of the idea. The text (Baxter and Rennie, 1996) and the on-line lecture notes of Professor Robert Kohn at www.math.nyu.edu/faculty/kohn/ are good places to learn more.

EXERCISES

12.1. ★★★ Using a large sheet of paper and a pen with plenty of ink, show that for $\mu = r$ the quantity $W(S, t)$ in (12.4) satisfies the Black–Scholes PDE (8.15). (You may differentiate inside the integral sign without worrying about whether this is justified.)

12.2. ★★★ Consider a European-style option with payoff at expiry given by $\Lambda(S(T)) = S(T)$. Explain why the time-zero value of this option must be S_0. By using (6.11), show that asking for the discounted expected payoff (12.1) to match this value leads immediately to the risk neutrality condition $\mu = r$.

12.3. ★★ Given initial asset price S_0 at time $t = 0$, show that, in a risk-neutral world, the factor $N(d_2)$ in the Black–Scholes formula (8.19) represents the probability that a European call option will be exercised.

12.4. ★★★ Show that the value $W(S, t)$ in (12.4) can be computed from the following recipe.

    (i)   Compute the Black–Scholes option value at (S, t) with the interest rate set to $r = \mu$.

    (ii) Scale this quantity by $e^{(\mu-r)(T-t)}$.

(This recipe was used to create Figure 12.1.)

12.5. ★★★ Consider the following, simplified scenario for valuing a European-style option.

- The time-zero asset price is S_0.
- At expiry, the asset price may take only two possible values

$$S(T) = S_{\text{up}} > S_0, \qquad \text{with probability } p,$$
$$S(T) = S_{\text{down}} < S_0, \qquad \text{with probability } 1 - p.$$

Let Λ denote the payoff function, and let $\Lambda_{\text{up}} := \Lambda(S_{\text{up}})$ and $\Lambda_{\text{down}} := \Lambda(S_{\text{down}})$ denote the two possible payoffs at expiry. Take a portfolio at time $t = 0$ consisting of A units of asset and an amount C of cash. Asking for this portfolio to *replicate the option* (i.e. to have payoff Λ_{up} when $S(T) = S_{\text{up}}$ and Λ_{down} when $S(T) = S_{\text{down}}$) leads to a pair of linear equations for A and C. Find and solve these to obtain

$$A = \frac{\Lambda_{\text{up}} - \Lambda_{\text{down}}}{S_{\text{up}} - S_{\text{down}}}, \tag{12.5}$$

$$C = e^{-rT} \left[\Lambda_{\text{down}} - \left(\frac{\Lambda_{\text{up}} - \Lambda_{\text{down}}}{S_{\text{up}} - S_{\text{down}}} \right) S_{\text{down}} \right]. \tag{12.6}$$

Then use the no arbitrage principle to deduce that a fair time-zero value for the option is

$$S_0 \left(\frac{\Lambda_{\text{up}} - \Lambda_{\text{down}}}{S_{\text{up}} - S_{\text{down}}} \right) + e^{-rT} \left(\frac{S_{\text{up}} \Lambda_{\text{down}} - S_{\text{down}} \Lambda_{\text{up}}}{S_{\text{up}} - S_{\text{down}}} \right). \tag{12.7}$$

Now, let

$$q := \frac{S_0 e^{rT} - S_{\text{down}}}{S_{\text{up}} - S_{\text{down}}}.$$

Use the no arbitrage principle to argue that $0 < q < 1$ must hold. Show that the value in (12.7) may also be interpreted as the discounted expected payoff of an asset taking the values

$$S(T) = S_{\text{up}} > S_0, \qquad \text{with probability } q,$$
$$S(T) = S_{\text{down}} < S_0, \qquad \text{with probability } 1 - q.$$

Can you see any features from this simplified scenario that carry through to the Black–Scholes version?

12.6. ★ ★ ★ In Section 10.3 we gave a financial interpretation of the inequality $\rho > 0$. Use the risk neutrality viewpoint to give an alternative interpretation.

12.5 Program of Chapter 12 and walkthrough

The program ch12, listed in Figure 12.2, illustrates risk neutrality in the manner of Figure 12.1. We fix S,E,r,sigma and T and an array of 200 values for mu. A for loop is then used to compute an array epayoff which stores the discounted time-zero Black–Scholes value when r is set to each mu value; see Exercise 12.4. This is done via the ch08 function from Chapter 8. After executing this loop, we use ch08 to obtain the true Black–Scholes value, C. We then plot the (muvals,epayoff) curve and superimpose a dashed line at height C.

PROGRAMMING EXERCISES

P12.1. Confirm experimentally the result mentioned in Exercise 12.3. Do this by generating a large number of expiry-time asset prices, and counting the proportion that are in-the-money.

P12.2. Investigate the use of quad and quadl for evaluating integrals of the form (12.4).

```
%CH12    Program for Chapter 12
%
%    Compute expected payoff for European call
%    Illustrates risk neutrality

clf

%%%%% Problem parameters %%%%%%
S = 5;  E = 7;  r = 0.08;  sigma = 0.3;  T = 1;
M = 200; muvals = linspace(0,0.16,M);
%%%%%%%%%%%%%%%%%%%%%%%%%%%%%%

epayoff = zeros(M,1);
for k = 1:M
    mu = muvals(k);
    % work out time-zero Black-Scholes value with r = mu
    [C, Cdelta, P, Pdelta] = ch08(S,E,mu,sigma,T);
    epayoff(k) = exp((mu-r)*T)*C;
end

% true Black–Scholes value
[C, Cdelta, P, Pdelta] = ch08(S,E,r,sigma,T);
plot(muvals,epayoff,'r-');
hold on, grid on
plot([muvals(1),muvals(end)],[C,C],'b-');
xlabel('\mu'), legend('Expected payoff','Black-Scholes')
```

Fig. 12.2.   Program of Chapter 12: ch12.m.

Quotes

... risk-neutrality is far from easy to grasp intuitively,
which is perhaps the source of the confusion above.
The key steps in the derivation of the Black–Scholes equation,
namely no arbitrage and that risk-free portfolios can earn the risk-free rate,
are intuitively clear.

    PAUL WILMOTT, SAM HOWISON AND JEFF DEWYNNE (Wilmott *et al.*, 1995)

Risk neutral valuation, which was developed by John Cox and Stephen Ross,
has the dual virtues that it can be applied to practically any option valuation problem
and it is marvelously intuitive.

    MARK P. KRITZMAN (Kritzman, 2000)

To put it simply,
if there is an arbitrage price, any other price is too dangerous to quote.

    MARTIN BAXTER AND ANDREW RENNIE (Baxter and Rennie, 1996)

13

Solving a nonlinear equation

OUTLINE

- general problem
- bisection method
- Newton's method

13.1 Motivation

In the next chapter, where we look at computing the implied volatility, we will need an algorithm for solving a nonlinear equation. This chapter introduces two such algorithms.

13.2 General problem

The task that we consider in this chapter is

given a function $F : \mathbb{R} \to \mathbb{R}$, find an $x^\star \in \mathbb{R}$ such that $F(x^\star) = 0$.

In general, of course, we cannot find an $x^\star$ analytically, and must therefore content ourselves with an approximation via a computational method. It is also worth keeping in mind that, depending on the nature of F, there may be no suitable $x^\star$, exactly one $x^\star$ or many $x^\star$ values.

13.3 Bisection

The *bisection method* is based on the observation that if a continuous function changes sign then it must pass through zero; that is,

for continuous F, if $x_a < x_b$ with $F(x_a)F(x_b) < 0$,
then $F(x^\star) = 0$ for some $x_a < x^\star < x_b$.

Having found x_a and x_b with $F(x_a)F(x_b) < 0$, we could evaluate F at the midpoint $x_{\mathrm{mid}} := (x_a + x_b)/2$. The sign of $F(x_{\mathrm{mid}})$ must then match either $F(x_a)$ or

123

$F(x_b)$. This means that one of the intervals $[x_a, x_{mid}]$ or $[x_{mid}, x_b]$ must contain an $x^\star$. By repeating this process, we can construct an arbitrarily small interval in which an $x^\star$ must lie – hence we can find an $x^\star$ to any level of accuracy.

We may thus spell out the bisection method as follows.

Step 1: Find x_a and x_b with $x_a < x_b$ such that $F(x_a)F(x_b) \leq 0$.
Step 2: Set $x_{mid} := (x_a + x_b)/2$ and evaluate $F(x_{mid})$.
Step 3: If $F(x_a)F(x_{mid}) < 0$ then reset $x_b = x_{mid}$.
            Otherwise reset $x_a = x_{mid}$.
Step 4: If $x_b - x_a < \varepsilon$ then stop. Use $\frac{1}{2}(x_a + x_b)$ as the approximation to $x^\star$.
            Otherwise return to Step 2.

Note that we must choose a value $\varepsilon > 0$ for our stopping criterion $x_b - x_a < \varepsilon$. It is easy to see that the value $(x_a + x_b)/2$ on termination is no more than a distance $\varepsilon/2$ from a solution $x^\star$. Hence, ε controls the accuracy of the process.

There is no foolproof procedure for finding suitable x_a and x_b in Step 1. Without specific knowledge of the function F we must resort to trial and error.

Because the bisection method halves the length of the interval $[x_a, x_b]$ on each iteration, we may bound the error at the kth iteration by $L/2^{k+1}$, where L is the length of the original interval, $x_b - x_a$. This is referred to as a *linear convergence bound* because the error bound decreases by a linear factor, in this case $\frac{1}{2}$, on each iteration. We consider next a faster method.

13.4 Newton

Newton's method (also called the *Newton–Raphson method*) can be derived in a number of ways. We will use a Taylor series approach. Suppose we wish to compute a sequence $x_0, x_1, x_2, \ldots$ that converges to a solution $x^\star$. We may expand $F(x_n + \delta)$ for small δ by

$$F(x_n + \delta) = F(x_n) + \delta F'(x_n) + O(\delta^2). \tag{13.1}$$

Ignoring the $O(\delta^2)$ term and setting $F(x_n) + \delta F'(x_n) = 0$ gives $\delta = -F(x_n)/F'(x_n)$. It follows that if x_n is close to a solution $x^\star$ then

$$x_{n+1} = x_n - \frac{F(x_n)}{F'(x_n)} \tag{13.2}$$

should be even closer. Given a starting value, x_0, the iteration (13.2) defines Newton's method.

Since we discarded an $O(\delta^2)$ term in (13.1), we may expect that the error

$x_n - x^\star$ squares as n increases to $n + 1$; that is, if $x_n - x^\star = O(\delta)$ then $x_{n+1} - x^\star = O(\delta^2)$. To see this more clearly, note that, using $F(x^\star) = 0$ and assuming $F'(x_n) \neq 0$ in (13.2), a Taylor series gives

$$x_{n+1} - x^\star = x_n - x^\star - \left(\frac{F(x_n) - F(x^\star)}{F'(x_n)} \right)$$

$$= x_n - x^\star - \frac{(x_n - x^\star)F'(x_n) + O\left((x_n - x^\star)^2\right)}{F'(x_n)}$$

$$= O\left((x_n - x^\star)^2\right). \tag{13.3}$$

This type of analysis can be formalized to give the following result.

Theorem 1 Suppose F has a continuous second derivative, and suppose $x^\star \in \mathbb{R}$ satisfies $F(x^\star) = 0$ and $F'(x^\star) \neq 0$. Then there exists a $\delta > 0$ such that for $|x_0 - x^\star| < \delta$ the sequence given by (13.2) is well defined for all $n > 0$,

$$\lim_{n \to \infty} |x_n - x^\star| = 0$$

and there exists a constant C such that

$$|x_{n+1} - x^\star| \leq C|x_n - x^\star|^2. \tag{13.4}$$

$\square$

The bound (13.4) shows that Newton's method has *quadratic* or *second order* convergence. However, the result requires the starting value x_0 to be chosen sufficiently close to $x^\star$. In practice Newton's method works very well when a suitable x_0 is found, but may fail to converge otherwise.

Computational example Suppose we wish to find the value of $x^\star$ such that $\mathbb{P}(X \leq x^\star) = \frac{2}{3}$, where $X \sim N(0, 1)$. Equivalently, we want to solve $F(x) = 0$, where $F(x) := N(x) - \frac{2}{3}$ with $N(x)$ defined in (3.18). It follows from the definition of $N(x)$ that $F(x)$ is an increasing function of x with $F(0) = \frac{1}{2} - \frac{2}{3} < 0$ and $\lim_{x \to \infty} F(x) = 1 - \frac{2}{3} > 0$. Hence, we may immediately conclude that $F(x) = 0$ has a unique solution $0 < x^\star < \infty$. This can be confirmed from the plot of $F(x)$ in Figure 13.1. We may apply the bisection method with $x_a = 0$ and with x_b sufficiently large that $F(x_b) > 0$. For the choice $x_b = 10$ and a tolerance of $\varepsilon = 10^{-5}$ in the stopping criterion, the errors $|x_{\mathrm{mid}} - x^\star|$ are shown as asterisks in the left-hand plot of Figure 13.2. Note that the y-axis is logarithmically scaled. We see that 20 iterations were taken in the bisection method. The dashed line corresponding to $10 \times \left(\frac{1}{2}\right)^{k+1}$ has been added to the plot. The preceding analysis shows that the error lies below this line. The right-hand plot in Figure 13.2 shows the corresponding errors for Newton's method. Here we set

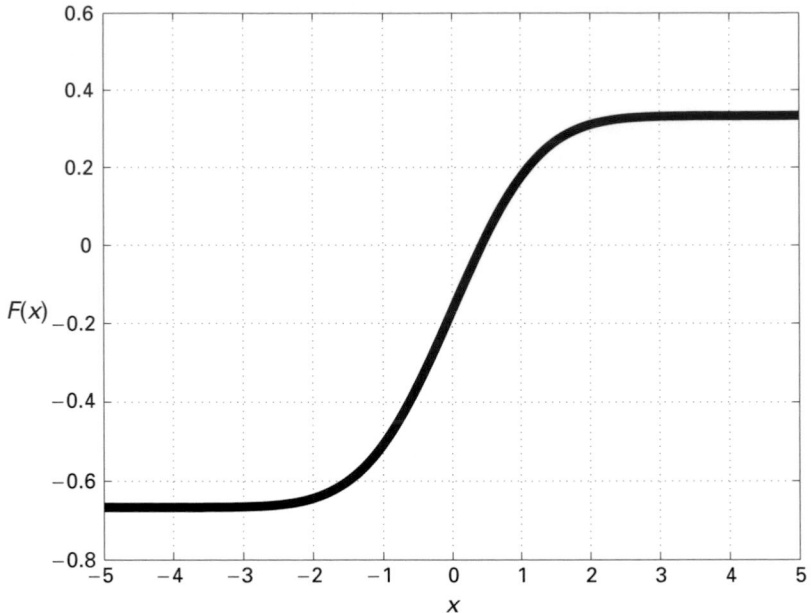

Fig. 13.1.   The function $F(x) := N(x) - \frac{2}{3}$.

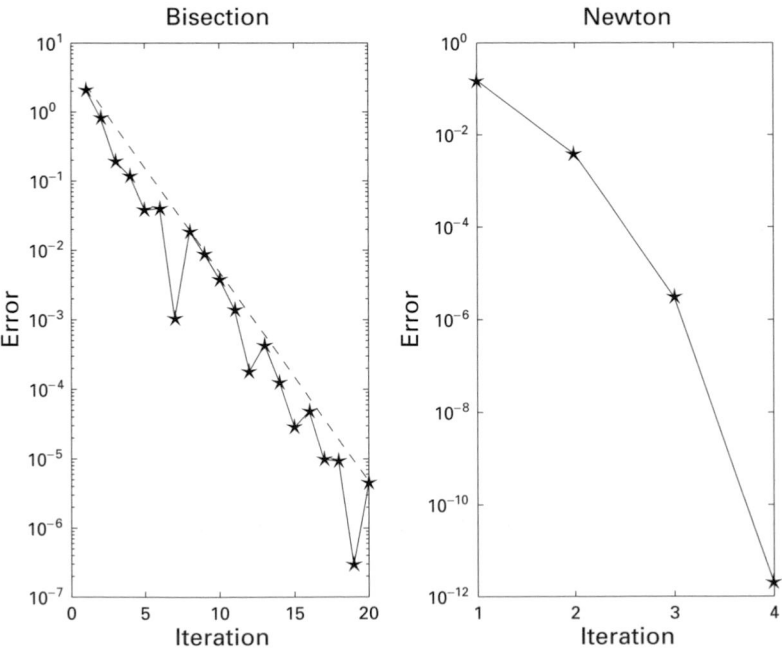

Fig. 13.2.   Error in the bisection method (left) and Newton's method (right). A reference line of slope -1 has been added in the left-hand plot.

$x_0 = 1$ and stopped when $|x_{n+1} - x_n| < 10^{-5}$. We see that only 4 iterations were required to produce an error of around 10^{-12}, and the error roughly squares from one step to the next. Repeating Newton's method with $x_0 = 2$, however, resulted in a sequence that 'blew up' – the numbers became too large for the computer to store. ◇

13.5 Further practical issues

There are many issues that we have not addressed here. It is possible, for example, to design a hybrid algorithm that uses a safe method, like bisection, until the iterates are close to an $x^\star$ and then switches to Newton's method to get the benefit of rapid convergence. Also, the residual $|F(x_n)|$ gives a measure of how close x_n is to a solution, and this can be incorporated into the stopping criterion. Furthermore, although we have considered only a single nonlinear equation, it is possible to generalize Newton's method to the case of many equations in many unknowns.

13.6 Notes and references

Most introductory numerical analysis texts have a chapter on solving nonlinear equations. An excellent and up-to-date specialist treatment that includes MAT-LAB codes is (Kelley, 1995). The classic advanced text is (Ortega and Rheinboldt, 1970).

If you need to brush up on Taylor series, order notation and, for the next chapter, the Mean Value Theorem, there are many introductory texts to choose from; (Estep, 2002) is an excellent modern treatment.

EXERCISES

13.1. ⋆ Suppose that Step 1 of the bisection method has been completed for a continuous function F and let $L = x_b - x_a$. In terms of L and ε, how many iterations of Steps 2–4 will be taken? Check that your answer is consistent with the left-hand plot in Figure 13.2.

13.2. ⋆⋆ Consider the following approach to computing a sequence of approximations $x_0, x_1, x_2, \ldots$ to $x^\star$. Given x_n, let x_{n+1} be the solution to $p_n(x) = 0$, where $p_n(x)$ is an approximation to $F(x)$ determined by the three conditions (a) $p_n(x)$ is linear, (b) $p_n(x_n) = F(x_n)$ and (c) $p'_n(x) = F'(x_n)$. Draw a picture to illustrate this construction and then show that x_{n+1} is given by (13.2). (Hence, this is an alternative derivation of Newton's method.)

13.3. ★★ To compute the errors that are shown in Figure 13.2 it was necessary to obtain the exact solution x^*. This was done by setting `xstar = sqrt(2)*erfinv(1/3)` where `erfinv` is MATLAB's built-in routine to evaluate the inverse error function described in Exercise 4.3. Confirm that `xstar` is the required solution.

13.4. ★★ Look at Figure 13.1. Using a ruler and pencil, and following the linearization approach in Exercise 13.2, convince yourself that Newton's method will converge with the starting value $x_0 = 1$, but will not converge with the starting value $x_0 = 2$.

13.7 Program of Chapter 13 and walkthrough

In ch13, listed in Figure 13.3, we apply Newton's method to $N(x) + e^x = 2$. The line

```
    exact = fzero(inline('0.5*(1+erf(x/sqrt(2))) + exp(x)- 2'),1);
```

```
%CH13      Program for Chapter 13
%
% Apply Netwon's method to N(x) + exp(x) = 2.

exact = fzero(inline('0.5*(1+erf(x/sqrt(2))) + exp(x)- 2'),1);

x0 = 1;
x = x0;
xdiff = 1;
k = 1;
kmax = 100;
tol = 1e-8;

while (xdiff >= tol & k < kmax)
    Fval = 0.5*(1+erf(x/sqrt(2))) + exp(x) - 2;
    Fprime = exp(-0.5*x^2)/sqrt(2*pi) + exp(x);
    increment = Fval/Fprime;
    x = x - increment;
    xnewton(k) = x;
    newterr(k) = abs(xnewton(k)-exact);
    k = k+1;
    xdiff = abs(increment);
end

format short e  % non-default for number display
disp('Newton error')
disp(newterr')
format % reset to default for number display
```

Fig. 13.3. Program of Chapter 13: `ch13.m`.

uses MATLAB's built-in equation solver `fzero` to compute an 'exact' solution, which we use for reference. The syntax

```
while (xdiff >= tol & k < kmax)
    .
    .
    .
    end
```

sets up a loop that repeats while **both** `xdiff >= tol` **and** `k < kmax` remain true. In other words, the loop terminates when **either** `xdiff` drops below `tol` **or** the maximum number, `kmax`, of iterations has been reached. Inside the loop we implement Newton's method for the problem. The error in each iterate is stored in the array `newterr`.

On exiting the loop, we output the errors. The line `format short e` sets up a number display format that is appropriate for this output. At the end of the program we reset the display to the default with `format`.

Output from `ch13` is

```
        Newton error
          1.5465e-01
          8.3622e-03
          2.4964e-05
          2.2279e-10
          1.1102e-16
```

This is consistent with the quadratic convergence discussed in Section 13.4 – the error roughly squares from one iteration to the next until it reaches a level that the machine cannot distinguish from zero.

PROGRAMMING EXERCISES

P13.1. Investigate the convergence of the bisection method on the problem solved by `ch13`.

P13.2. Using your answer to programming exercise P12.2, apply bisection to confirm that the two curves displayed in Figure 12.1 intersect at $\mu = r$.

Quotes

Chance has put in our way a most singular and whimsical problem,
and its solution is its own reward.

> SHERLOCK HOLMES, in *The Adventure of the Blue Carbuncle* by Sir Arthur Conan
> Doyle

A blunder is an accidental mistake,
as opposed to an approximation error, which is merely a compromise.

> ROBERT M. CORLESS (Corless, 2002)

14

Implied volatility

14.1 Motivation

We now put the bisection method and Newton's method to work on the problem of computing the implied volatility.

14.2 Implied volatility

The Black–Scholes call and put values depend on S, E, r, $T - t$ and σ^2. Of these five quantities, only the asset volatility σ cannot be observed directly. How do we find a suitable value for σ? One approach is to extract the volatility from the observed market data – given a quoted option value, and knowing S, t, E, r and T, find the σ that leads to this value. Having found σ, we may use the Black–Scholes formula to value other options on the same asset. A σ computed this way is known as an *implied volatility*. The name indicates that σ is implied by option value data in the market.

A completely different way to get hold of σ is described in Chapter 20.

We focus here on the case of extracting σ from a European call option quote. An analogous treatment can be given for a put, or, alternatively, the put quote could be converted into a call quote via put–call parity (8.23).

14.3 Option value as a function of volatility

We assume that the parameters E, r and T and the asset price S and time t are known. (In practice, we will typically be interested in the time-zero case, $t = 0$

131

and $S = S_0$.) We thus treat the option value as a function of σ only, and, for the rest of this chapter, denote it by $C(\sigma)$. Given a quoted value $C^\star$, our task is to find the implied volatility $\sigma^\star$ that solves $C(\sigma) = C^\star$.

Computing the implied volatility requires the solution of a nonlinear equation and hence, from Chapter 13, we may use the bisection method or Newton's method. We will find that it is possible to exploit the special form of the nonlinear equation arising in this context.

Since volatility is non-negative, only values $\sigma \in [0, \infty)$ are of interest. Let us look at $C(\sigma)$ in the case of large or small volatility. First, as $\sigma \to \infty$, we see from (8.20) that $d_1 \to \infty$ and hence $N(d_1) \to 1$. Similarly, from (8.21), as $\sigma \to \infty$, $d_2 \to -\infty$ and hence $N(d_2) \to 0$. It follows in (8.19) that

$$\lim_{\sigma \to \infty} C(\sigma) = S. \tag{14.1}$$

Next, we look at the limit $\sigma \to 0^+$ and separate out three cases.

Case 1: $S - Ee^{-r(T-t)} > 0$. In this case $\log(S/E) + r(T-t) > 0$, so as $\sigma \to 0^+$ we have $d_1 \to \infty$, $N(d_1) \to 1$, $d_2 \to \infty$ and $N(d_2) \to 1$. Hence, $C \to S - Ee^{-r(T-t)}$.

Case 2: $S - Ee^{-r(T-t)} < 0$. In this case $\log(S/E) + r(T-t) < 0$, so as $\sigma \to 0^+$ we have $d_1 \to -\infty$, $N(d_1) \to 0$, $d_2 \to -\infty$ and $N(d_2) \to 0$. Hence, $C \to 0$.

Case 3: $S - Ee^{-r(T-t)} = 0$. In this case $\log(S/E) + r(T-t) = 0$, so as $\sigma \to 0^+$ we have $d_1 \to 0$, $N(d_1) \to \frac{1}{2}$, $d_2 \to 0$ and $N(d_2) \to \frac{1}{2}$. Hence, $C \to \frac{1}{2}(S - Ee^{-r(T-t)}) = 0$.

The three cases are summarized neatly by the formula

$$\lim_{\sigma \to 0^+} C(\sigma) = \max(S - Ee^{-r(T-t)}, 0). \tag{14.2}$$

Now we recall from Chapter 10 that the derivative of C with respect to σ, that is, the vega, is given by (10.6). In particular, we know that $\partial C/\partial \sigma > 0$. Since $C(\sigma)$ is continuous with a positive first derivative, we conclude that C is monotonic increasing on $[0, \infty)$. From (14.1) and (14.2), values of $C(\sigma)$ must lie between $\max(0, S - Ee^{-r(T-t)})$ and S. It follows that $C(\sigma) = C^\star$ has a solution if and only if

$$\max(S - Ee^{-r(T-t)}, 0) \leq C^\star < S, \tag{14.3}$$

and if a solution exists it is unique. Henceforth, we assume that this condition holds. For further justification of this assumption we note from Section 2.6 that if (14.3) is violated then an arbitrage opportunity exists.

For later use, we will calculate the second derivative. Differentiating (10.6) gives

$$\frac{\partial^2 C}{\partial \sigma^2} = -\frac{S\sqrt{T-t}}{\sqrt{2\pi}} e^{-\frac{1}{2}d_1^2} d_1 \frac{\partial d_1}{\partial \sigma}.$$

From (8.20) we have

$$\frac{\partial d_1}{\partial \sigma} = -\frac{\log(S/E) + r(T-t)}{\sigma^2 \sqrt{T-t}} + \frac{1}{2}\sqrt{T-t}$$

$$= -\left[\frac{\log(S/E) + (r - (\sigma^2/2))(T-t)}{\sigma^2 \sqrt{T-t}}\right]$$

$$= -\frac{d_2}{\sigma}$$

and hence

$$\frac{\partial^2 C}{\partial \sigma^2} = \frac{S\sqrt{T-t}}{\sqrt{2\pi}} e^{-\frac{1}{2}d_1^2} \frac{d_1 d_2}{\sigma} = \frac{d_1 d_2}{\sigma} \frac{\partial C}{\partial \sigma}. \tag{14.4}$$

It follows from (14.4) that $\partial C/\partial \sigma$ is maximum over $[0, \infty)$ at $\sigma = \widehat{\sigma}$, where

$$\widehat{\sigma} := \sqrt{2 \left| \frac{\log S/E + r(T-t)}{T-t} \right|}, \tag{14.5}$$

see Exercise 14.1. Moreover, $\partial^2 C/\partial \sigma^2$ may be written in the form

$$\frac{\partial^2 C}{\partial \sigma^2} = \frac{T-t}{4\sigma^3} (\widehat{\sigma}^4 - \sigma^4) \frac{\partial C}{\partial \sigma}, \tag{14.6}$$

see Exercise 14.2. The identity (14.6) shows us that $C(\sigma)$ is *convex* for $\sigma < \widehat{\sigma}$ and *concave* for $\sigma > \widehat{\sigma}$. This will allow us to get a globally convergent Newton iteration by suitably choosing the starting value.

14.4 Bisection and Newton

We will write our nonlinear equation for $\sigma^\star$ in the form $F(\sigma) = 0$, where $F(\sigma) := C(\sigma) - C^\star$. To apply the bisection method, we require an interval $[\sigma_a, \sigma_b]$ over which $F(\sigma)$ changes sign. It follows from (14.1), (14.2) and the monotonicity of $C(\sigma)$ that this can be done by fixing K (say $K = 0.05$) and trying $[0, K]$, $[K, 2K]$, $[2K, 3K], \ldots$.

Newton's method takes the form

$$\sigma_{n+1} = \sigma_n - \frac{F(\sigma_n)}{F'(\sigma_n)}, \tag{14.7}$$

where $F'(\sigma) = \partial C / \partial \sigma$ is given by (10.6). Because we know a lot about F, we can exploit an expansion along the lines of (13.3) that keeps track of the remainder. Using $F(\sigma^\star) = 0$ and the Mean Value Theorem, we have

$$\sigma_{n+1} - \sigma^\star = \sigma_n - \sigma^\star - \frac{F(\sigma_n) - F(\sigma^\star)}{F'(\sigma_n)}$$

$$= \sigma_n - \sigma^\star - \frac{(\sigma_n - \sigma^\star) F'(\xi_n)}{F'(\sigma_n)},$$

for some ξ_n between σ_n and $\sigma^\star$. Hence, we may write

$$\frac{\sigma_{n+1} - \sigma^\star}{\sigma_n - \sigma^\star} = 1 - \frac{F'(\xi_n)}{F'(\sigma_n)}. \tag{14.8}$$

We know that $F'(\sigma)$ is positive and takes its maximum at the point $\widehat{\sigma}$ in (14.5). Hence, using the starting value $\sigma_0 = \widehat{\sigma}$ we must have $0 < F'(\xi_0) < F'(\widehat{\sigma})$ in (14.8), so that

$$0 < \frac{\sigma_1 - \sigma^\star}{\sigma_0 - \sigma^\star} < 1. \tag{14.9}$$

This means that the error in σ_1 is smaller than, but has the same sign as, the error in σ_0. To proceed we suppose that $\widehat{\sigma} < \sigma^\star$. Then (14.9) tells us that $\sigma_0 < \sigma_1 < \sigma^\star$. Now, we know from (14.6) that $F''(\sigma) < 0$ for all $\sigma > \widehat{\sigma}$ and we also know that ξ_1 in (14.8) lies between σ_1 and $\sigma^\star$. Hence $0 < F'(\xi_1) < F'(\sigma_1)$ and (14.8) gives

$$0 < \frac{\sigma_2 - \sigma^\star}{\sigma_1 - \sigma^\star} < 1.$$

Continuing this argument gives

$$0 < \frac{\sigma_{n+1} - \sigma^\star}{\sigma_n - \sigma^\star} < 1, \qquad \text{for all } n \geq 0. \tag{14.10}$$

So the error decreases monotonically as n increases.

In a similar manner, it can be shown that (14.10) holds in the case where $\widehat{\sigma} > \sigma^\star$, see Exercise 14.3. Overall, we conclude that with the choice $\sigma_0 = \widehat{\sigma}$ the error will always decrease monotonically as n increases. It follows that the error must tend to zero, and the theory from Chapter 13 then shows that convergence must be quadratic. Hence, $\sigma_0 = \widehat{\sigma}$ is a foolproof starting value for Newton's method on this particular nonlinear equation. This is therefore our method of choice for computing the implied volatility.

Computational example Figure 14.1 illustrates the performance of Newton's method in the case where $S_0 = 3$, $E = 1$, $r = 0.05$, $T = 3$ and $t = 0$. We used

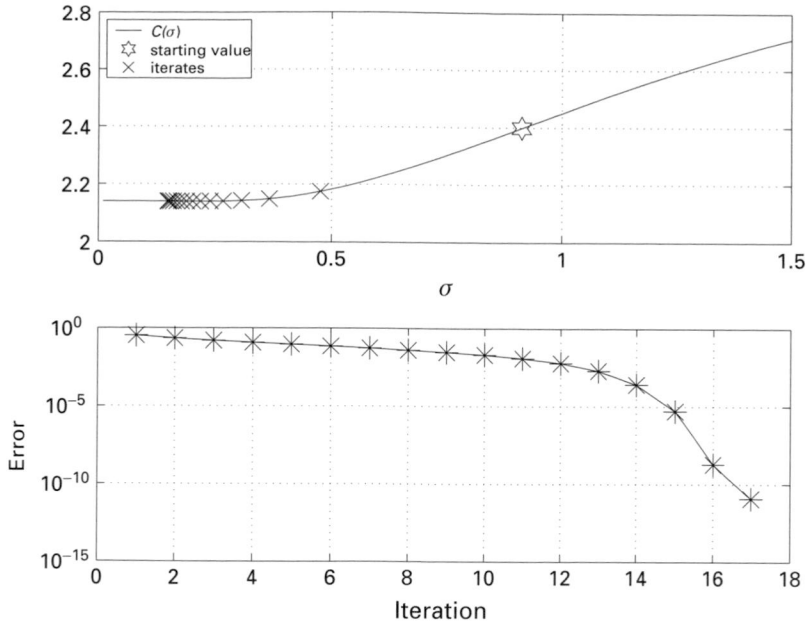

Fig. 14.1. Newton's method for the implied volatility. Upper picture: iterates. Lower picture: errors.

$\sigma^\star = 0.15$ in order to compute the Black–Scholes value for C, and then applied Newton's method to see how quickly $\sigma^\star$ could be found. We took the starting value $\sigma_0 = \widehat{\sigma}$ given by (14.5), so monotonic convergence is guaranteed. The upper picture in Figure 14.1 shows the curve $C(\sigma)$ and superimposes the starting value $(\sigma_0, C(\sigma_0))$ and the subsequent iterates $(\sigma_n, C(\sigma_n))$. The lower picture plots the size of the error, $|\sigma_n - \sigma^\star|$. We see that the initial convergence is quite slow, but ultimately the characteristic second order behaviour emerges. The slow initial decrease in the error may be caused by the fact that $F'(\sigma^\star)$ is close to zero. (Recall that $F'(\sigma^\star) \neq 0$ is an assumption in the convergence theorem. If $F'(\sigma^\star)$ were exactly zero then Newton's method would converge at a rate slower than quadratic – see (Ortega and Rheinboldt, 1970), for example.) ◇

14.5 Implied volatility with real data

We now look at the implied volatility for call options traded on the London International Financial Futures and Options Exchange (LIFFE), as reported in the *Financial Times* on Wednesday, 22 August 2001. The data is for the FTSE 100 index, which is an average of 100 equity shares quoted on the London Stock Exchange. The expiry date for these options was December 2001.

Exercise price	Option price
5125	475
5225	405
5325	340
5425	$280\frac{1}{2}$
5525	226
5625	$179\frac{1}{2}$
5725	139
5825	105

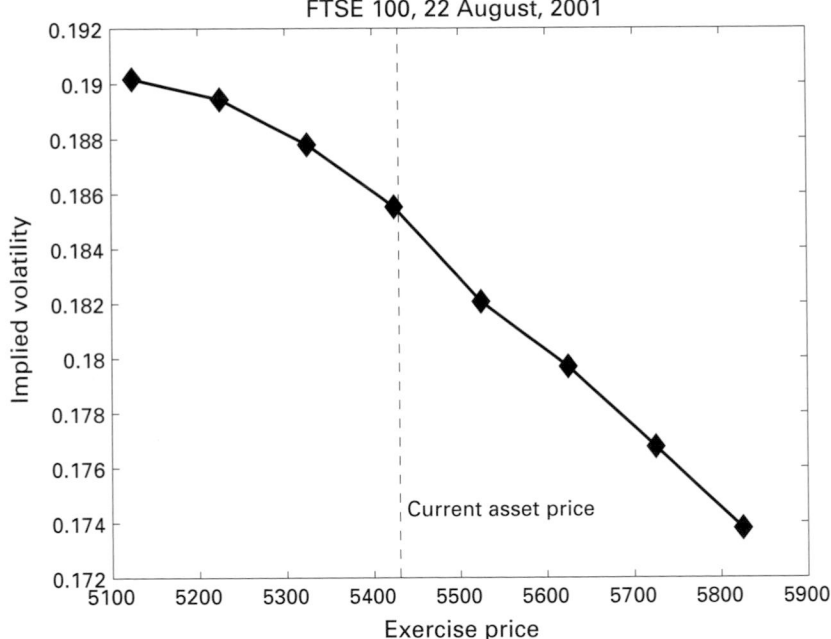

Fig. 14.2. Implied volatility against exercise price for some FTSE 100 index data.

The initial asset price (on 22 August 2001) was 5430.3. We took values of $r = 0.05$ for the interest rate and $T = 4/12$ for the duration of the option. Figure 14.2 shows the implied volatility computed for the eight different exercise prices. Of course, if the Black–Scholes formula were valid, the volatility would be the same for each exercise price. We see that in this example the implied volatility varies by around 10%. We also note that the implied volatility is higher for options that start in-the-money than for options starting out-of-the-money. This behaviour is typical for data arising after the stock market crash of October 1987. Pre-crash plots of implied volatility against exercise price would often produce a convex *smile* shape; more recent data tends to produce more of a *frown*.

14.6 Notes and references

The convergence analysis for Newton's method is based on the article (Manaster and Koehler, 1982). It is also mentioned in (Kwok, 1998). More about implied volatility can be found in (Hull, 2000; Kwok, 1998), for example.

The widely reported phenomenon that the implied volatility is not constant as other parameters are varied does, of course, imply that the Black–Scholes formulas fail to describe perfectly the option values that arise in the marketplace. This should be no surprise, given that the theory is based on a number of simplifying assumptions. Despite the disparities, the Black–Scholes theory, and the insights that it provides, continue to be regarded highly by both academics and market traders. Indeed, it is common for option values to be quoted in terms of 'vol'; rather than giving $C^\star$, the $\sigma^\star$ such that $C(\sigma^\star) = C^\star$ in the Black–Scholes formula is used to describe the value.

Many attempts have been made to 'fix' the nonconstant volatility discrepancy in the Black–Scholes theory. A few of these have met with some success, but none lead to the simple formulas and clean interpretations of the original work. Chapter 17 of (Hull, 2000) gives a good overview of the directions that have been taken.

EXERCISES

14.1. ★★★ Show that $\partial C / \partial \sigma$ has a unique maximum over $[0, \infty)$ at $\sigma = \widehat{\sigma}$, where $\widehat{\sigma}$ is defined in (14.5).

14.2. ★★ Verify the identity (14.6).

14.3. ★★★ Suppose $\sigma_0 = \widehat{\sigma}$. Using the fact that $F''(\sigma) > 0$ for $\sigma < \widehat{\sigma}$, confirm that (14.10) holds in the case where $\widehat{\sigma} > \sigma^\star$.

14.7 Program of Chapter 14 and walkthrough

In ch14, listed in Figure 14.3, we implement Newton's method for implied volatility of a European call. After setting up r,S,E,T and tau, we use ch08 from Chapter 8 to compute the call value, C_true, corresponding to a volatility of sigma_true=0.3. Our task is then to recover the volatility that produces the call value C_true. We use a while loop of the form discussed for ch13, with a call to ch10 providing the required vega value. The final solution is correct to within 6×10^{-17}.

```
%CH14 Program for Chapter 14
%
% Computes implied volatility for a European call

%%%%%%%%%%% parameters %%%%%%%%%%%
r = 0.03; S = 2; E = 2; T = 3; tau = T; sigma_true = 0.3;
[C_true, Cdelta, P, Pdelta] = ch08(S,E,r,sigma_true,tau);
%%%%%%%%%%%%%%%%%%%%%%%%%%%%%%%%%%%

%starting value
sigmahat = sqrt(2*abs( (log(S/E) + r*T)/T ) );

%%%%% Newton's method %%%%%
tol = 1e-8;
sigma = sigmahat;
sigmadiff = 1;
k = 1;
kmax = 100;
while (sigmadiff >= tol & k < kmax)
   [C, Cdelta, Cvega, P, Pdelta, Pvega] = ch10(S,E,r,sigma,tau);
   increment = (C-C_true)/Cvega;
   sigma = sigma - increment;
   k = k+1;
   sigmadiff = abs(increment);
end
sigma
```

Fig. 14.3.   Program of Chapter 14: ch14.m.

PROGRAMMING EXERCISES

P14.1. Alter ch14 to deal with a put option.

P14.2. Acquire some real option data, either electronically or via a newspaper, and create a figure like Figure 14.2. If possible, investigate the behaviour of the implied volatility as the expiry time varies.

Quotes

The volatility is the most important and elusive quantity
in the theory of derivatives.

PAUL WILMOTT (Wilmott, 1998)

A smiley implied volatility is the wrong number
to put in the wrong formula
to obtain the right price.

RICCARDO REBONATO (Rebonato, 1999)

It is the strong opinion of the author that most traders

will gain an improved performance by concentrating their efforts
on a better prediction of the volatility input into a Black–Scholes type model
rather than introducing other pricing techniques.

A. L. H. SMITH (Smith, 1986)

In those days, before the publication of the Black–Scholes option-pricing formula,
warrants were often grossly mispriced. Thorpe soon developed a computer program
to identify such opportunities; its deployment was so successful that,
by 1970, both Thorpe and Kassouf had abandoned academe for greener pastures.

JAMES CASE, reviewing the book (Bass, 1999) in *Society for Industrial and Applied Mathematics (SIAM) News*, Jan/Feb, 2001.

15

Monte Carlo method

15.1 Motivation

Chapter 12 showed that valuing an option can be regarded as computing an expected value. The idea of using pseudo-random number generators to compute estimates of expected values was touched on in Chapter 4. Here we pull these two threads together and introduce the Monte Carlo approach to valuing an option. As we will see in Chapter 19, this provides a powerful means to compute option values in cases where no analytical formulas are available.

15.2 Monte Carlo

To begin, we consider the case of a general random variable X, whose expected value $\mathbb{E}(X) = a$ and variance $\text{var}(X) = b^2$ are not known. Suppose

- we are interested in computing an approximation to a (and possibly b), and
- we are able to take independent samples of X using a pseudo-random number generator.

We know from Table 4.2 that computing the average of a large number of samples can give a good approximation to the mean. Hence, if we let $X_1, X_2, \ldots, X_M$ denote independent random variables with the same distribution as X then we might expect

$$a_M := \frac{1}{M} \sum_{i=1}^{M} X_i \qquad (15.1)$$

141

to be a good approximation to a. We say that an approximation to $\mathbb{E}(X)$ is *un-biased* if it has the same expected value as X. It is easily shown that a_M in (15.1) is unbiased; see Exercise 15.1. To estimate the variance, since $\text{var}(X) := \mathbb{E}((X - \mathbb{E}(X))^2)$, an obvious choice is $(\sum_{i=1}^{M}(X_i - a_M)^2)/M$. However, to make this estimate unbiased we need to re-scale it slightly. Exercise 15.2 asks you to check that the appropriate unbiased version is

$$b_M^2 := \frac{1}{M-1} \sum_{i=1}^{M} (X_i - a_M)^2. \tag{15.2}$$

By the Central Limit Theorem, $\sum_{i=1}^{M} X_i$ behaves like an $\mathsf{N}(Ma, Mb^2)$ random variable, so

$$a_M - a \text{ is approximately } \mathsf{N}\left(0, \frac{b^2}{M}\right). \tag{15.3}$$

We could also say that $a_M - a$ is approximately an $\mathsf{N}(0, 1)$ random variable scaled by $b/\sqrt{M}$. This suggests that sampling a_M for large M should give an approximation to a that is correct to $O(1/\sqrt{M})$.

We can make this argument more quantitative by using the idea of a confidence interval that was introduced in Section 6.5. If we had equality in (15.3) then, from (6.15),

$$\mathbb{P}\left(a - \frac{1.96\,b}{\sqrt{M}} \leq a_M \leq a + \frac{1.96\,b}{\sqrt{M}}\right) = 0.95.$$

We may re-write this as

$$\mathbb{P}\left(a_M - \frac{1.96\,b}{\sqrt{M}} \leq a \leq a_M + \frac{1.96\,b}{\sqrt{M}}\right) = 0.95. \tag{15.4}$$

The ratio $b/\sqrt{M}$ appearing in (15.4) is often refered to as the *standard error*. Replacing the unknown b by the approximation b_M we see that the unknown expected value a lies in the interval

$$\left[a_M - \frac{1.96\,b_M}{\sqrt{M}}, a_M + \frac{1.96\,b_M}{\sqrt{M}}\right] \tag{15.5}$$

with probability 0.95, approximately. In other words (15.5) gives an approximate *95% confidence interval* for a.

This analysis leads us to the basic Monte Carlo method for approximating a. We compute M independent samples and form a_M in (15.1). In order to monitor the error, we also compute the variance approximation b_M^2 in (15.2). Having b_M allows us to compute the confidence interval (15.5) (or indeed, a confidence interval for some other percentage, such as 99%; see Exercise 6.8).

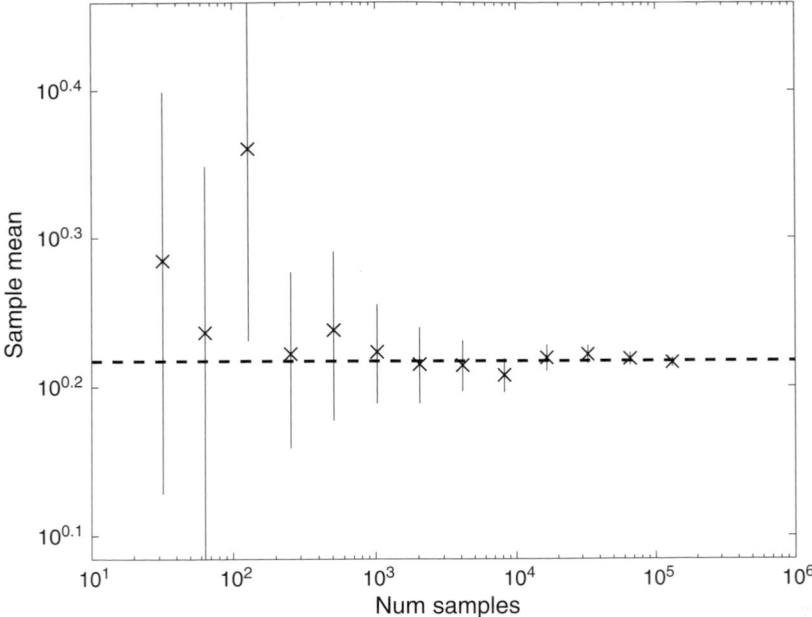

Fig. 15.1. Monte Carlo approximations to $\mathbb{E}(e^Z)$, where $Z \sim N(0, 1)$. Crosses are the approximations, vertical lines give computed 95% confidence intervals. Horizontal dashed line is at height $\mathbb{E}(e^Z) = \sqrt{e}$.

There are two key features to note.

(i) The size of the confidence interval shrinks like the *inverse square root* of the number of samples. To reduce the 'error' by a factor of 10 requires a hundredfold increase in the sample size. This is a severe limitation that typically makes it impossible to get very high accuracy from a Monte Carlo approximation.

(ii) The size of the confidence interval is directly proportional to the standard deviation, that is the square root of the variance, of the random variable under consideration. In practice, it is highly desirable to transform the problem of approximating $\mathbb{E}(X)$ to the problem of approximating $\mathbb{E}(Y)$ where Y is another random variable that has the same mean as X but a smaller variance. This idea, known as *variance reduction*, forms a vital part of practical Monte Carlo algorithms. The two most popular approaches are covered in Chapters 21 and 22.

Computational example In Figure 15.1 we give results from a Monte Carlo simulation of $\mathbb{E}(e^Z)$, where $Z \sim N(0, 1)$. In this case we can work out analytically that $\mathbb{E}(e^Z) = \sqrt{e}$; see Exercise 15.3. We used 13 different sample sizes, $M = 2^5, 2^6, 2^7, \ldots, 2^{17}$. For each sample size, the picture plots the computed mean, a_M, with a cross and gives the computed 95% confidence interval as a vertical line, often called an *error bar*. Note that both axes have logarithmic scales. The exact mean, $\sqrt{e}$, is represented as a dashed line. We see that as M increases the computed mean generally becomes more accurate and the confidence interval shrinks. In the third case, $M = 2^7$, the correct mean is not contained in the confidence interval. Remember that our theory predicts that this will happen roughly 5% of the time, but requires M sufficiently large that

- the Central Limit Theorem approximation is accurate, and
- the computed variance b_M approximates well the exact variance b.

A separate check revealed that with $M = 2^7$ the variance error $|b_M^2 - b^2|$ was a non-negligible 7.1. The errors for $M = 2^{16}$ and $M = 2^{17}$ are 5.31×10^{-3} and 3.64×10^{-3}, respectively. The ratio of these errors is ≈ 1.5, which is close to the asymptotic $(M \to \infty)$ value of $\sqrt{2}$. This computation is typical – we have achieved a few digits of accuracy with a modest amount of work. The 'curse of the $1/\sqrt{M}$' makes higher accuracy extremely costly. To reduce the error to, say, 10^{-4} would take of the order of 10^8 samples, and to reduce it to 10^{-6} would take of the order of 10^{12} samples; see Exercise 15.4.   ◇

15.3 Monte Carlo for option valuation

We are now in a position to use Monte Carlo for option valuation. We consider a European-style option with payoff that is some function Λ of the asset price at expiry. Our model for the asset price at expiry is (6.8) with $t = T$. Using the risk neutrality approach discussed in Chapter 12, the time-zero option value can be found by setting $\mu = r$ and computing $e^{-rT} \mathbb{E}(\Lambda(S(T)))$.

Putting all this together, we wish to find the expected value of the random variable

$$ e^{-rT} \Lambda \left(S_0 \exp \left[\left(r - \frac{1}{2}\sigma^2 \right) T + \sigma \sqrt{T} Z \right] \right), \qquad \text{where } Z \sim \mathsf{N}(0, 1). \quad (15.6) $$

The resulting Monte Carlo algorithm can be summarized as follows:

> for $i = 1$ to M
>   compute an $\mathsf{N}(0, 1)$ sample ξ_i
>   set $S_i = S_0 e^{(r - \frac{1}{2}\sigma^2)T + \sigma\sqrt{T}\xi_i}$
>   set $V_i = e^{-rT} \Lambda(S_i)$
> end
> set $a_M = \left(\frac{1}{M}\right) \sum_{i=1}^{M} V_i$
> set $b_M^2 = \left(\frac{1}{(M-1)}\right) \sum_{i=1}^{M} (V_i - a_M)^2$

The output provides an approximate option price a_M and an approximate 95% confidence interval (15.5).

Computational example We now use the Monte Carlo method to value a European call option, so $\Lambda(S(T)) = \max(S(T) - E, 0)$. We will use the Black–Scholes formula (8.19) to compute the exact value and then see how

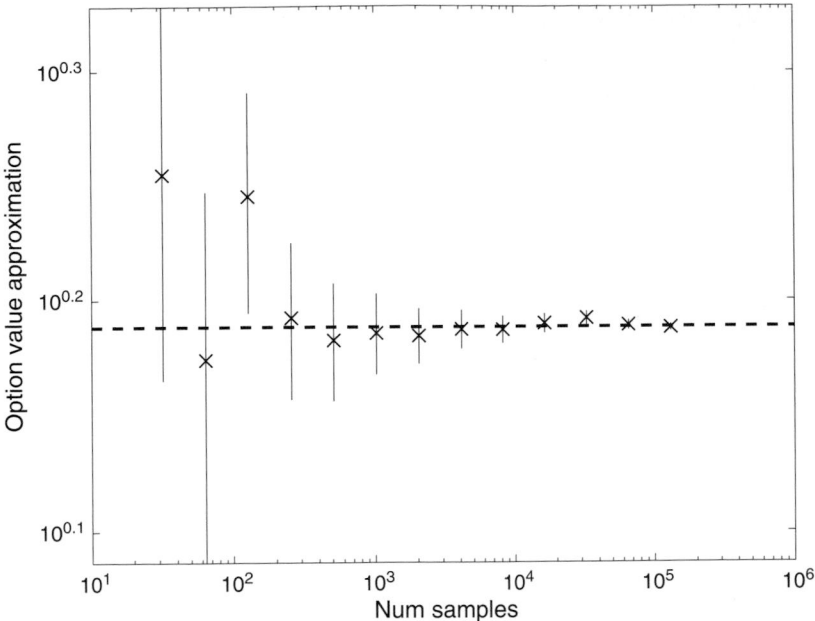

Fig. 15.2. Monte Carlo approximations to a European call option value. Crosses are the approximations, vertical lines give computed 95% confidence intervals. Horizontal dashed line is at height given by the Black–Scholes formula.

well Monte Carlo performs. We take $S_0 = 10$, $E = 9$, $\sigma = 0.1$, $r = 0.06$ and $T = 1$. The Black–Scholes option value is 1.5429. Figure 15.2 shows the Monte Carlo results, in a similar manner to Figure 15.1. We used sample sizes $M = 2^5, 2^6, \ldots, 2^{17}$. For each sample size we plot the computed mean a_M with a cross and show the computed 95% confidence interval as a vertical line. The same pseudo-random number sequences as those for Figure 15.1 were used, and once again the $M = 2^7$ confidence interval does not contain the true mean. With the largest sample size, 2^{17}, the error in a_M was $\approx 1.2 \times 10^{-3}$. We emphasize that in this example there is no need to apply Monte Carlo as the Black–Scholes formula gives the exact solution conveniently. However, as we will see in Chapter 19, Monte Carlo comes into its own in more complicated circumstances where no simple Black–Scholes-type formula is available. $\diamond$

15.4 Monte Carlo for Greeks

In addition to the option value, we know that the Greeks – the partial derivatives of the option value with respect to various quantities – are also of interest. In particular, the delta, $\Delta := \partial V / \partial S$, plays a key role in the hedging strategy that a trader must operate in order to replicate the option. The Monte Carlo approach can be used to compute approximate partial derivatives; although it must be handled with care and may prove expensive. We focus here on the case of computing the time-zero delta of a European-style option with payoff $\Lambda(S(T))$, but the principles apply generally.

A simple Taylor series expansion shows that the delta at asset price S and time t satisfies

$$\frac{\partial V(S,t)}{\partial S} = \frac{V(S+h,t) - V(S,t)}{h} + O(h), \qquad \text{as } h \to 0. \qquad (15.7)$$

Hence, we may choose a small value of h and use the *finite difference* approximation

$$\frac{\partial V(S,t)}{\partial S} \approx \frac{V(S+h,t) - V(S,t)}{h}.$$

This produces an approximation to delta at a single point based on option values at two points with slightly different S arguments. Using the risk-neutral, discounted, expected payoff formulation, we could thus approximate the time-zero delta by computing Monte Carlo estimates of the two expected values in

$$e^{-rT} \frac{\mathbb{E}\left(\Lambda(S(T))\right), \text{ with } S(0) = S_0 + h, \text{ with } S(0) = S_0) - \mathbb{E}\left(\Lambda(S(T))\right)}{h}.$$

$$(15.8)$$

Now the error in each of the two Monte Carlo estimates is $O(1/\sqrt{M})$ and hence, after dividing the difference by h, we expect an overall error of $O(1/(h\sqrt{M}))$ for (15.8). This is unfortunate: we want to make h small to get a good derivative approximation in (15.7), but doing so forces us to take even more samples than the basic Monte Carlo option value strategy would need. Another way to view the difficulty is to note that in order to satisfy ourselves that we have even got the correct sign for the delta, we might ask for non-overlapping confidence intervals from the two Monte Carlo approximations. Since the exact means $V(S,t)$ and $V(S+h,t)$ differ by $O(h)$, this requires confidence interval widths that are at least as small as $O(h)$.

However, we can claw back some accuracy by noting that the two random variables in (15.8) are highly *correlated*: for any particular asset path the payoff starting from $S(0) = S_0$ is likely to be close to the payoff starting from $S_0 + h$. Intuitively, the corresponding sample mean errors should be close *provided that we use the same paths for the two simulations*; that is we apply Monte Carlo to the equivalent problem

$$e^{-rT} \frac{\mathbb{E}\left[(\Lambda(S(T))), \text{ with } S(0) = S_0 + h) - (\Lambda(S(T))), \text{ with } S(0) = S_0)\right]}{h},$$

$$(15.9)$$

which involves a single random variable. In Chapters 21 and 22 we make the idea of correlation more explicit, and Exercise 22.3 gives further justification for this argument.

This leads us to the following Monte Carlo algorithm for approximating Δ.

> for $i = 1$ to M
> > compute an $N(0, 1)$ sample ξ_i
> > set $S_i = S_0 e^{(r-\frac{1}{2}\sigma^2)T+\sigma\sqrt{T}\xi_i}$
> > set $S_i^h = (S_0 + h)e^{(r-\frac{1}{2}\sigma^2)T+\sigma\sqrt{T}\xi_i}$
> > set $\Delta_i = e^{-rT}(\Lambda(S_i^h) - \Lambda(S_i))/h$
> end
> set $a_M = \frac{1}{M}\sum_{i=1}^{M}\Delta_i$
> set $b_M^2 = \frac{1}{M-1}\sum_{i=1}^{M}(\Delta_i - a_M)^2$

This produces an approximate delta value a_M and an approximate 95% confidence interval (15.5).

Computational example Here we return to the European call option used for Figure 15.2 with the same sample sizes M. The Black–Scholes time-zero delta value is 0.9558. Figure 15.3 shows the corresponding delta approximations from

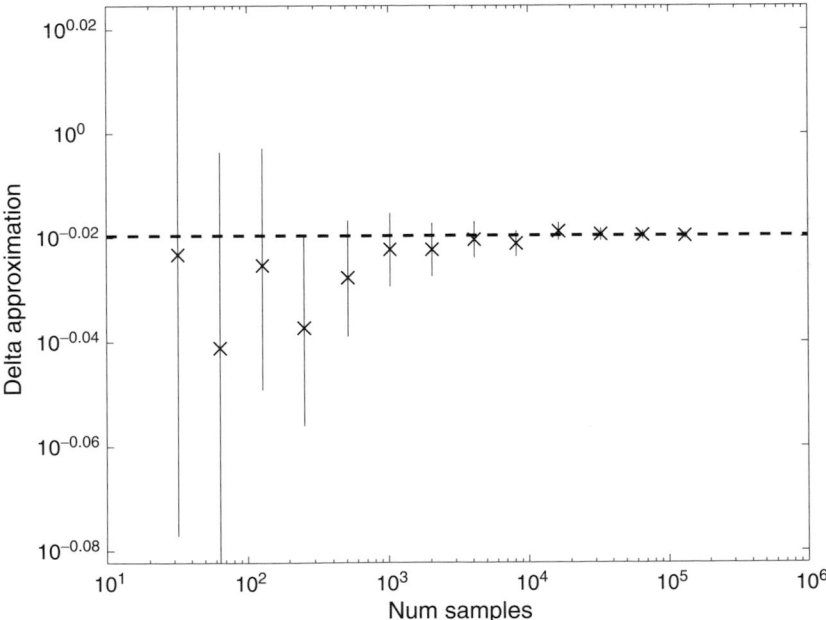

Fig. 15.3. Monte Carlo approximations to time-zero delta of a European call option. Crosses are the approximations, vertical lines give computed 95% confidence intervals. Horizontal dashed line is at height given by the Black–Scholes formula.

the algorithm above, in the style of Figures 15.1 and 15.2. We fixed $h = 10^{-4}$. For $M = 2^{17}$, the error in the sample was $\approx 1.2 \times 10^{-4}$; smaller than that for the corresponding Monte Carlo option value approximation. We also experimented with the corresponding algorithm that uses *different* pseudo-random numbers for the two options. The results were much worse; for the M values used here, no digits of accuracy were recorded, and the standard deviations were around $50\,000$ times larger.   ◇

15.5 Notes and references

There are many texts that discuss general Monte Carlo simulation. A 'golden oldie' that is still highly relevant is (Hammersley and Handscombe, 1964), whilst a short and very accessible modern perspective is given by (Madras, 2002). Monte Carlo, pseudo-random number generation and other simulation issues are treated in detail in (Ripley, 1987).

Boyle's classic 1977 paper (Boyle, 1977), which won the *Journal of Financial Economics'* All-Star Paper Award 2002, introduced Monte Carlo for option valuation. The paper (Boyle *et al.*, 1997) summarizes developments since then, and in particular, has a detailed treatment of the Greeks. Texts that cover Monte Carlo for finance in some depth include (Clewlow and Strickland, 1998; Jäckel, 2002; Kwok, 1998)

EXERCISES

15.1. ⋆ Show that a_M in (15.1) is an unbiased estimator of $\mathbb{E}(X)$; that is, $\mathbb{E}(a_M) = a$.

15.2. ⋆ ⋆ ⋆ Show that

$$\widehat{b}^2_M := \frac{1}{M} \sum_{i=1}^{M} (X_i - a_M)^2$$

satisfies

$$\mathbb{E}(\widehat{b}^2_M) = \left(\frac{M-1}{M} \right) b^2. \tag{15.10}$$

This confirms that $\widehat{b}_M{}^2$ is not an unbiased estimator of $\mathsf{var}(X)$. Conclude from (15.10) that $b_M{}^2$ in (15.2) **is** an unbiased estimator of $\mathsf{var}(X)$.

15.3. ⋆⋆ Show that if $Z \sim \mathsf{N}(0, 1)$ then $\mathbb{E}(e^Z) = \sqrt{e}$. [Hint: recall (3.8).]

15.4. ⋆⋆ For the computational experiment that produced Figure 15.1, it was predicted that 'To reduce the error to, say, 10^{-4}, would take of the order of 10^8

```
%CH15     Program for Chapter 15
%
%   Monte Carlo for a European put

randn('state',100)

%%%%%%%%%%% Problem and method parameters %%%%%%%%%%
S = 4; E = 5; sigma = 0.3; r = 0.04; T = 1;
Dt = 1e-3; N = T/Dt; M = 1e4;
%%%%%%%%%%%%%%%%%%%%%%%%%%%%%%%%%%%%%%%%%%%%%%%%%

V = zeros(M,1);
for i = 1:M
   Sfinal = S*exp((r-0.5*sigma^2)*T+sigma*sqrt(T)*randn);
   V(i) = exp(-r*T)*max(E-Sfinal,0);
end
aM = mean(V); bM = std(V);
conf = [aM - 1.96*bM/sqrt(M), aM + 1.96*bM/sqrt(M)]
```

Fig. 15.4.   Program of Chapter 15: ch15.m.

samples, and to reduce it to 10^{-6} would take of the order of 10^{12} samples.' Where do these figures come from? For the computations in Figure 15.2, roughly how many samples would be needed to reduce the error to 10^{-6}?

15.6 Program of Chapter 15 and walkthrough

In ch15, listed in Figure 15.4, we use Monte Carlo to value a European put. The code follows the algorithm in Section 15.3, making use of MATLAB's built-in functions mean and std, which, respectively, compute the sample mean (15.1) and sample standard deviation – the square root of the sample variance (15.2).

The code produces a confidence interval conf = [1.0070, 1.0402]. Checking with the Black–Scholes formula from ch08 gives

```
>> [C, Cdelta, P, Pdelta] = ch08(4,5,0.04,0.3,1)
   C = 0.2167
   Cdelta = 0.3226
   P = 1.0207
   Pdelta = -0.6774
```

PROGRAMMING EXERCISES

P15.1. Adapt ch15 to produce a picture like that in Figure 15.2.

P15.2. Adapt ch15 to produce an estimate of the delta.

Quotes

To know the vintage and quality of a wine
one need not drink the whole cask.

OSCAR WILDE, 1854–1900.

In the classical theory, which we are discussing here,
the unknown parameter p is a *number*, not a random variable,
so p is either in I or outside it,
and it is meaningless to speak of the probability of p lying in I
(the Bayesians, on the other hand, consider p a random variable – see Section 15.7).
The expression *95% confidence interval* refers to the
procedure through which I was produced.
This procedure produces intervals containing p 95 percent of the time.

RICHARD ISAAC (Isaac, 1995)

The Central Limit Theorem is a powerful tool,
and we wish we had an intuitive explanation of why it should be true.
Unfortunately, we don't.

MARK DENNEY AND STEVEN GAINES (Denney and Gaines, 2000)

16

Binomial method

16.1 Motivation

We now introduce another computational approach. The binomial method is straightforward to describe and implement, and, as we will see in Chapters 18 and 19, has the advantage that it is readily adapted to a range of non-European options for which no analytical formula is available. In particular, the binomial method provides the simplest means to value American options. In studying the method, we revisit two ideas, discrete asset price models and risk neutrality.

16.2 Method

The binomial method uses a simple discrete model for the asset price movement. We let $\delta t = T/M$ denote the spacing between successive time points, where T is the expiry date. So asset prices will be considered at times $t_i = i\delta t$, for $0 \leq i \leq M$. A key assumption in the binomial method is that between successive time levels the asset price moves either up by a factor u or down by a factor d. An upward movement occurs with probability p and a downward movement occurs with probability $1 - p$. This scenario can be regarded as a simplified version of the discrete model introduced in Chapter 6. Indeed, Exercise 16.1 asks you to cast this simple model in the form (6.2) by redefining Y_i.

Since the initial asset price, S_0, is known, at time $t_1 = \delta t$ the possible asset prices are uS_0 and dS_0. Similarly, at time $t_2 = 2\delta t$ the possible asset prices are

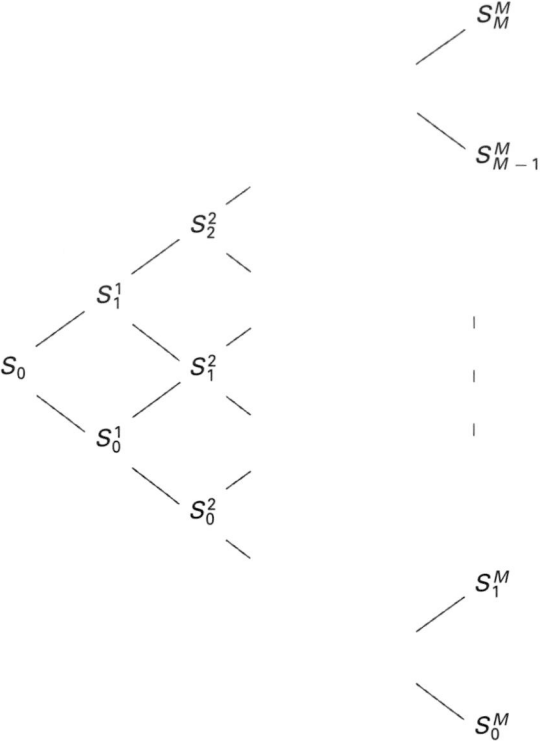

Fig. 16.1.   Recombining binary tree of asset prices.

$u^2 S_0$, $ud S_0$ and $d^2 S_0$. (The price $ud S_0$ may arise from an upward movement followed by a downward movement or from a downward movement followed by an upward movement.) In general, at time $t = t_i := i \delta t$ there are $i + 1$ possible asset prices, which we label

$$S_n^i = d^{i-n} u^n S_0, \qquad 0 \le n \le i. \tag{16.1}$$

Hence, at the expiry time $t = t_M = T$ there are $M + 1$ possible asset prices. The values S_n^i for $0 \le n \le i$ and $0 \le i \le M$ form a recombining binary tree, as illustrated in Figure 16.1.

For a European-style call option, the payoff at expiry has the form $\Lambda(S(T))$. Hence, if the asset has price S_n^M at time $t = t_M = T$ then the value of the option at that time is $\Lambda(S_n^M)$. Generally, we let V_n^i denote the value of the option at time $t = t_i$ corresponding to asset price S_n^i. We thus know that

$$V_n^M = \Lambda(S_n^M), \qquad 0 \le n \le M. \tag{16.2}$$

Our task is to find V_0^0, the option value at time zero. We may do this by working backwards through the tree. Suppose $\{V_n^{i+1}\}_{n=0}^{i+1}$ are known; that is, we have the option values corresponding to time $t = t_{i+1}$ and all possible asset prices. Then consider the option value V_n^i corresponding to asset price S_n^i at time $t = t_i$. Because of our up/down assumption about the asset price movement, working from right to left, the asset price S_n^i comes either from S_{n+1}^{i+1}, with probability p, or from S_n^{i+1}, with probability $1 - p$. Now, recall the definition (3.1) for the expected value of a discrete random variable. The big idea in the binomial method is to multiply the two possible values V_{n+1}^{i+1} and V_n^{i+1} by their associated probabilities to get an expected value. In this way the option value V_n^i corresponding to asset price S_n^i is taken to be $pV_{n+1}^{i+1} + (1 - p)V_n^{i+1}$, scaled by the appropriate factor that allows for the interest rate, r. This gives the fundamental relation

$$V_n^i = e^{-r\delta t}\left(pV_{n+1}^{i+1} + (1 - p)V_n^{i+1}\right), \qquad 0 \le n \le i, \qquad 0 \le i \le M - 1.$$
$$(16.3)$$

Once the parameters u, d, p and M have been chosen, the formulas (16.1)–(16.3) completely specify the binomial method. The recurrence (16.1) shows how to insert the asset prices in the binomial tree. Having obtained the asset prices at time $t = t_M = T$, (16.2) gives the corresponding option values at that time. The relation (16.3) may then be used to step backwards through the tree until V_0^0, the option value at time $t = t_0 = 0$, is computed.

16.3 Deriving the parameters

Since the discrete asset price model in the binomial method fits into the framework of (6.2), by appealing to Exercise 6.2 we could tune the parameters by asking for the corresponding Y_i to have zero mean and unit variance. This would lead to two constraints. However, to give more insight into the workings of the method, we will derive those constraints from first principles. Exercise 16.5 asks you to confirm that the two approaches lead to the same conclusion.

As a means to write down an expression for the up/down asset price model used in the binomial method, we define a random variable R_i such that $R_i = 1$ if the asset price goes up from time $(i - 1)\delta t$ to $i\delta t$ and $R_i = 0$ if the asset price goes down. Hence, $R_i = 1$ with probability p and $R_i = 0$ with probability $1 - p$. This means that R_i is a Bernoulli random variable with parameter p, so from (3.2) and (3.14) we see that $\mathbb{E}(R_i) = p$ and $\text{var}(R_i) = p(1 - p)$. After n time increments the asset has undergone $\sum_{i=1}^n R_i$ upward movements and $n - \sum_{i=1}^n R_i$ downward movements. So the asset price $S(n\delta t)$ at time $t = n\delta t$ is given by

$$S(n\delta t) = S_0 u^{\sum_{i=1}^n R_i} d^{n - \sum_{i=1}^n R_i}.$$

We may re-arrange this to

$$\frac{S(n\delta t)}{S_0} = d^n \left(\frac{u}{d}\right)^{\sum_{i=1}^{n} R_i}.$$

Taking logs gives

$$\log\left(\frac{S(n\delta t)}{S_0}\right) = n \log d + \log\left(\frac{u}{d}\right) \sum_{i=1}^{n} R_i. \tag{16.4}$$

Now, by the Central Limit Theorem, for large n the sum $\sum_{i=1}^{n} R_i$ behaves like a normal random variable. Hence, for large n, $\log(S(n\delta t)/S_0)$ will be close to normal. To match the continuous asset price model (6.8) used in the Black–Scholes analysis, we thus require the mean of $\log(S(n\delta t)/S_0)$ to be $(\mu - \frac{1}{2}\sigma^2)n\delta t$ and the variance to be $\sigma^2 n\delta t$. Further, as the binomial method works with expected values, we impose the risk neutrality assumption $\mu = r$. This leads to the conditions

$$p \log u + (1 - p) \log d = (r - \tfrac{1}{2}\sigma^2)\delta t, \tag{16.5}$$

$$\log\left(\frac{u}{d}\right) = \sigma \sqrt{\frac{\delta t}{p(1 - p)}}, \tag{16.6}$$

see Exercise 16.2. Regarding $\delta t = T/M$ as pre-specified, we now have two equations in the three unknowns, p, u and d. In general, we can fix one of the three and solve for the other two. To pick out a particular solution this way, we may set $p = \frac{1}{2}$ and solve to find that

$$u = e^{\sigma\sqrt{\delta t} + (r - \frac{1}{2}\sigma^2)\delta t}, \qquad d = e^{-\sigma\sqrt{\delta t} + (r - \frac{1}{2}\sigma^2)\delta t}, \tag{16.7}$$

see Exercise 16.3.

16.4 Binomial method in practice

The arguments in the previous section suggest that the binomial method asset model matches that used in the Black–Scholes analysis for small δt, that is, large M. We may thus hope that the option values computed from the binomial method agree well with those from the Black–Scholes formulas, and that the agreement improves if M is increased.

Computational example We use the binomial method to value a European put with $S_0 = 9$, $E = 10$, $T = 3$, $r = 0.06$ and $\sigma = 0.3$. Table 16.1 shows the results for $M = 100$, $M = 200$ and $M = 400$, along with the Black–Scholes value 1.4728. Our first observation is that with all three choices of M the binomial method approximation is correct to at least two decimal places. The

Table 16.1. *European put value*
approximations from binomial method

	Option value
$M = 100$	1.4716
$M = 200$	1.4762
$M = 400$	1.4726
Black–Scholes	1.4728

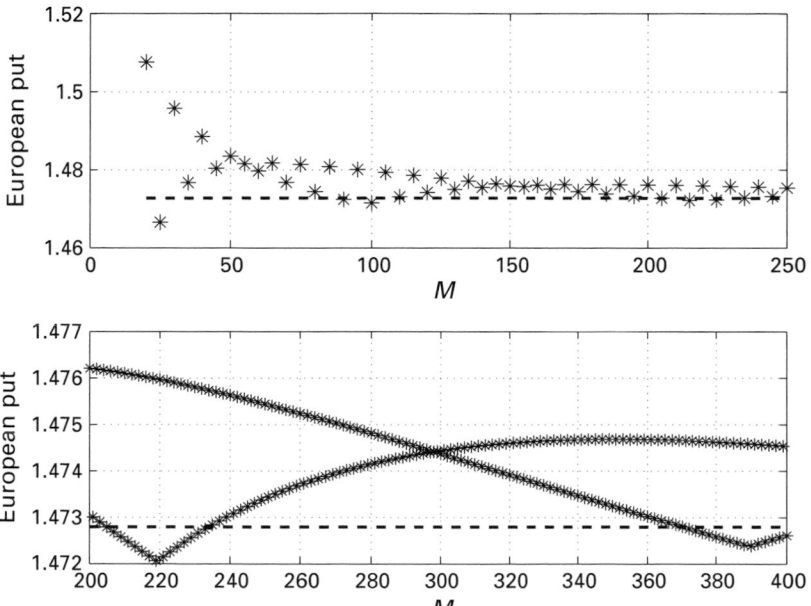

Fig. 16.2.　Convergence of the binomial method for a European put as the number of time points, M, increases. Upper picture: M goes from 20 to 250 in steps of 5. Dashed line is 'exact' solution. Lower picture: M goes from 200 to 400 in steps of 1.

most accurate approximation of the three comes from the largest value of M, which is intuitively reasonable. However, it is perhaps surprising that $M = 200$ gives less accuracy that $M = 100$. To check whether this is simply a quirk, the upper picture in Figure 16.2 shows the computed option value for $M = 20, 25, 30, \ldots, 250$, with the Black–Scholes value superimposed as a dashed line. We see that although the binomial method approximations do appear to converge as M increases, the convergence is by no means monotonic – taking a slightly bigger M may worsen the error – and there is a general 'sawtooth' pattern to the sequence of approximations as M increases. The lower plot emphasizes the waviness. Here we have plotted the computed solution for all M between 200 and 400. The result appears to oscillate between two smooth curves, neither of which approaches the correct answer monotonically. ◇

Two features stand out in Figure 16.2.

(i) The binomial method approximation converges to the Black–Scholes value as $M \rightarrow \infty$.
(ii) The convergence is not monotonic.

These may be shown to be generic. Moreover, it is possible to describe the rate at which convergence takes place. Letting $e_M = |V_0^0 - P(S_0, 0)|$ denote the error in the binomial method approximation with $\delta t = T/M$, it can be shown that there is a constant K such that

$$e_M \leq \frac{K}{M}. \tag{16.8}$$

In the upper picture of Figure 16.3 we display the errors in the example above for M between 100 and 400. The points have been joined by straight lines for clarity. The curve $1/M$ is added as a solid line, and we see that (16.8) appears to hold with $K = 1$. Taking logs in (16.8) gives $\log e_M \leq \log K - \log M$, showing that the log of the error as a function of $\log M$ should lie below a straight line of slope -1. The lower picture of Figure 16.3 re-scales the axes logarithmically to confirm this behaviour.

16.5 Notes and references

Cox, Ross and Rubinstein (Cox *et al.*, 1979) wrote the original binomial method paper. Since then numerous authors have analysed and extended the ideas.

It is possible to derive the parameters u, d and p from a number of different viewpoints. For example, with $p = \frac{1}{2}$ the choice

$$u = e^{r\delta t}\left(1 + \sqrt{e^{\sigma^2 \delta t} - 1}\right), \qquad d = e^{r\delta t}\left(1 - \sqrt{e^{\sigma^2 \delta t} - 1}\right) \tag{16.9}$$

is common; see (Kwok, 1998; Wilmott *et al.*, 1995). Exercise 16.4 shows that this is very close to the choice (16.7) for small δt.

Although much literature has been devoted to establishing that the error in various classes of binomial methods tends to zero as $M \rightarrow \infty$, surprisingly little attention was initially paid to the *rate* of convergence. Leisen and Reimer (Leisen and Reimer, 1996) developed a general convergence rate theory, and the bound (16.8) follows from their results. A more detailed analysis, with explicit error constants, appears in (Walsh, 2003).

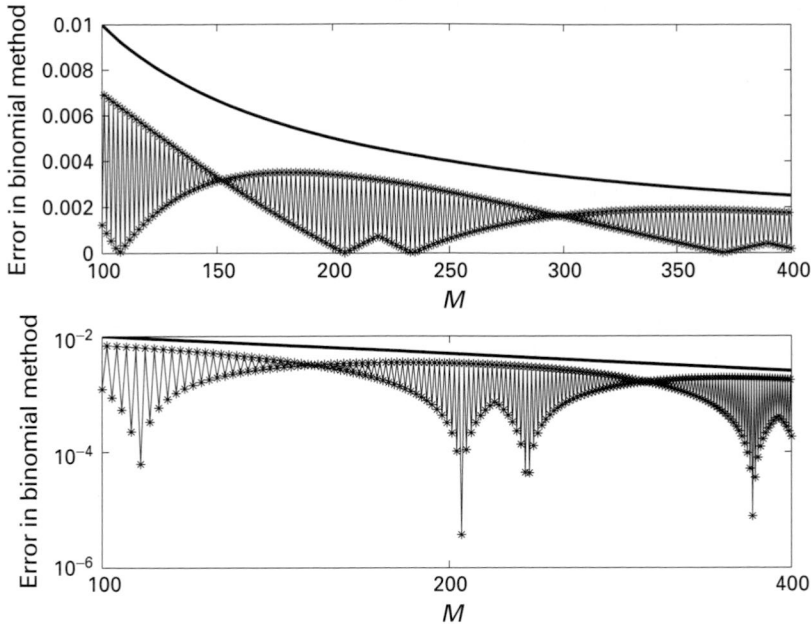

Fig. 16.3. Upper picture: Error in the binomial method for a European put as the number of time points, M, increases from 100 to 400. Solid line is $1/M$. Lower picture: same data on a log–log scale.

The odd–even ripples in the error, as depicted in Figures 16.2 and 16.3, have been widely reported. The references (Leisen and Reimer, 1996; Rogers and Stapleton, 1998) give explanations for the effect and propose fixes.

Applying the binomial method may be shown to be equivalent to using a finite difference method to approximate the Black–Scholes PDE, a point that we pursue in Section 24.4. This is one means of proving that the binomial method solution converges to the Black–Scholes value as $M \to \infty$, see (Kwok, 1998), for example, and numerical analysis insights can also be used to explain the odd-even ripples.

The book (Clewlow and Strickland, 1998) covers a number of practical issues in the implementation of the binomial method, and provides pseudo-code listings.

A case study with the aim of making the binomial method run as quickly as possible in MATLAB is given in (Higham, 2002), along with downloadable codes.

It is possible to compute Greeks via the binomial method. For partial derivatives with respect to S or t, approximations can be obtained using information from the tree. Exercise 16.8 illustrates the idea. Other partial derivatives can be treated by re-running the method with perturbed data, in the manner outlined in Section 15.4. Further details can be found in (Hull, 2000), for example, and (Walsh, 2003) shows that delta can be approximated to the same order of accuracy as the option value.

EXERCISES

16.1. ⋆ Consider the discrete asset price model used in the binomial method. Show that it may be written in the form (6.2) if we let Y_i be defined as

$$Y_i = \begin{cases} \frac{u-1-\mu\delta t}{\sigma\sqrt{\delta t}}, & \text{with probability } p, \\[2mm] \frac{d-1-\mu\delta t}{\sigma\sqrt{\delta t}}, & \text{with probability } 1-p. \end{cases} \tag{16.10}$$

16.2. ⋆⋆ Starting from (16.4) show that

$$\mathbb{E}\left[\log\left(\frac{S(n\delta t)}{S_0}\right)\right] = n\log d + \log\left(\frac{u}{d}\right)np$$

and

$$\text{var}\left[\log\left(\frac{S(n\delta t)}{S_0}\right)\right] = \left(\log\left(\frac{u}{d}\right)\right)^2 np(1-p).$$

Hence, obtain (16.5)–(16.6).

16.3. ⋆⋆ Show that setting $p = \frac{1}{2}$ in (16.5)–(16.6) produces (16.7).

16.4. ⋆⋆⋆ For the parameters u and d in (16.7) show that

$$u = 1 + \sigma\sqrt{\delta t} + r\delta t + O(\delta t^{3/2}), \qquad d = 1 - \sigma\sqrt{\delta t} + r\delta t + O(\delta t^{3/2}),$$

as $\delta t \to 0$.

Show also that the corresponding u and d parameters in (16.9) have the same expansions up to $O(\delta t^{3/2})$. [Hint: recall that $\sqrt{1+x} = 1 + \frac{1}{2}x + O(x^2)$ and $e^x = 1 + x + \frac{1}{2}x^2 + O(x^3)$ as $x \to 0$.]

16.5. ⋆⋆ We know from Exercise 6.2 that if Y_i in (16.10) has zero mean and unit variance, we recover the continuous asset price model in the limit $\delta t \to 0$. Set $\mu = r$ and $p = \frac{1}{2}$ and show that requiring $\mathbb{E}(Y_i) = 0$ and $\text{var}(Y_i) = 1$ in (16.10) leads to

$$u = 1 + \sigma\sqrt{\delta t} + r\delta t, \qquad d = 1 - \sigma\sqrt{\delta t} + r\delta t.$$

Note that these values agree with those in Exercise 16.4 up to $O(\delta t^{3/2})$.

16.6. ⋆⋆⋆ Returning to the recurrence (16.3) we see that for $M = 1$

$$V_0^0 = e^{-r\delta t}\left(pV_1^1 + (1-p)V_0^1\right),$$

and for $M = 2$

$$\begin{aligned} V_0^0 &= e^{-r\delta t}\left(pV_1^1 + (1-p)V_0^1\right) \\ &= e^{-r\delta t}\left(pe^{-r\delta t}(pV_2^2 + (1-p)V_1^2) + (1-p)e^{-r\delta t}(pV_1^2 \right. \\ &\qquad \left. + (1-p)V_0^2)\right) \\ &= e^{-2r\delta t}\left(p^2 V_2^2 + 2p(1-p)V_1^2 + (1-p)^2 V_0^2\right). \end{aligned}$$

Similarly for $M = 3$ we find that

$$V_0^0 = e^{-3r\delta t}\left(p^3 V_3^3 + 3p^2(1-p)V_2^3 + 3p(1-p)^2 V_1^3 + (1-p)^3 V_0^3\right).$$

The coefficients $\{1, 1\}$, $\{1, 2, 1\}$, $\{1, 3, 3, 1\}$ are familiar from Pascal's triangle. Having spotted this connection, prove by induction that

$$V_0^0 = e^{-rT} \sum_{k=0}^{M} \binom{M}{k} p^k (1-p)^{M-k} V_k^M, \qquad (16.11)$$

where $\binom{M}{k}$ denotes the binomial coefficient,

$$\binom{M}{k} := \frac{M!}{k!\,(M-k)!}.$$

16.7. ⋆⋆ Letting

$$W^i := \begin{bmatrix} V_0^i \\ V_1^i \\ \vdots \\ \vdots \\ V_i^i \end{bmatrix},$$

write down the form of the $(i + 1)$ by $(i + 2)$ matrix B^i such that $W^i = B^i W^{i+1}$.

16.8. ⋆⋆ Explain why the ratio $(V_1^1 - V_0^1)/(S_1^1 - S_0^1)$ can be regarded as an approximation to the time-zero delta.

16.6 Program of Chapter 16 and walkthrough

The program ch16 implements the binomial method for a European call. It is listed in Figure 16.4. First, parameters are initialized, using (16.7) for u and d. The quantity

```
S*d.^([M:-1:0]').*u.^([0:M]')
```

is an M+1 by 1 array whose components cover the values $S_0^M, S_1^M, \ldots, S_M^M$ in the expiry-time level of the asset price tree in Figure 16.1. Hence, the line

```
W = max(S*d.^([M:-1:0]').*u.^([0:M]')-E,0);
```

contains the expiry time option values, as in (16.2). We then work through the iteration (16.3) by exploiting MATLAB's colon notation to extract subarrays. The syntax

```
exp(-r*dt)*(p*W(2:i+1) + (1-p)*W(1:i));
```

```
%CH16        Program for Chapter 16
%
% Implements binomial method for European call

%%%%%%%% Problem and method parameters %%%%%%%%%%%%
S = 3; E = 2; T = 1; r = 0.05; sigma = 0.3;
M = 400; dt = T/M; p =0.5;
u = exp(sigma*sqrt(dt) + (r-0.5*sigma^2)*dt);
d = exp(-sigma*sqrt(dt) + (r-0.5*sigma^2)*dt);
%%%%%%%%%%%%%%%%%%%%%%%%%%%%%%%%%%%%%%%%%%%%%

% Time T option values
W = max(S*d.^([M:-1:0]').*u.^([0:M]')-E,0);

% Work back to option value at time zero
for i = M:-1:1
   W = exp(-r*dt)*(p*W(2:i+1) + (1-p)*W(1:i));
end

disp('Option value is'), disp(W)
```

Fig. 16.4.   Program of Chapter 16: ch16.m.

represents

$$
e^{-r\delta t} \left(p \begin{bmatrix} W_2 \\ W_3 \\ \vdots \\ W_{i+1} \end{bmatrix} + (1-p) \begin{bmatrix} W_1 \\ W_2 \\ \vdots \\ W_i \end{bmatrix} \right).
$$

The line for i = M:-1:1 sets up a loop that is repeated M times; first with i = M, then with i = M-1, and so on, down to i = 1. With this set-up, the dimension of W decreases by one each time around the loop. On exit, W is a scalar, whose value is V_0^0.

Running ch16.m produces the value 1.1175. To check, we may call ch08.

```
>> [C, Cdelta, P, Pdelta] = ch08(3,2,0.05,0.3,1)
   C = 1.1175
   Cdelta = 0.9524
   P = 0.0200
   Pdelta = -0.0476
```

PROGRAMMING EXERCISES

P16.1. Alter ch16 so that the choice (16.9) for *u* and *d* is used.

P16.2. Implement the binomial method via the formula (16.11).

Quotes

'Would you tell me, please, which way I ought to go from here?'
'That depends a good deal on where you want to get to,' said the Cat.
'I don't much care where . . . ' said Alice.
'Then it doesn't matter which way you go,' said the Cat.

LEWIS CARROLL, Alice in Wonderland

Sir, In your otherwise beautiful poem (*The Vision of Sin*)
there is a verse which reads
'Every moment dies a man,
every moment one is born.'
Obviously, this cannot be true and I suggest that in the next edition you have it read
'Every moment dies a man,
every moment $1\frac{1}{16}$ is born.'
Even this value is slightly in error
but should be sufficiently accurate for poetry.

CHARLES BABBAGE (in a letter to Lord Tennyson), source (Fröberg, 1985)

In the literature,
there are numerous contributions with limit proofs to European type options.
Astonishingly, however, the convergence speed of binomially computed option prices
has, so far, rarely been examined technically.
Here, we present a theorem . . .

DIETMAR LEISEN AND MATTHIAS REIMER (Leisen and Reimer, 1996)

17

Cash-or-nothing options

17.1 Motivation

We now take our first step away from vanilla Europeans and look at cash-or-nothing call and put options. There are three good reasons to look at these options.

- They are widely traded, and hence of practical importance.
- The corresponding Black–Scholes values can be found analytically.
- They give us another opportunity to investigate the risk neutrality idea.

17.2 Cash-or-nothing options

A *cash-or-nothing call option* differs from a European call option in that the payoff at expiry is

$$
\begin{array}{ll}
A, & \text{if } S(T) > E, \text{ and} \\
0, & \text{if } S(T) < E,
\end{array}
$$

where $A > 0$ is fixed. Holding this option amounts to making a straight bet that the terminal asset price will exceed the exercise price, E, that is, the European call will finish in-the-money. Winning the bet gets you A, losing the bet gets you nothing. Unlike the European case, there is no added value to be had from the asset exceeding the strike by a wide margin; the upside is limited to A.

We have not yet specified the payoff for the case $S(T) = E$. This is an exceptional event, technically it occurs with zero probability, so the resulting payoff is

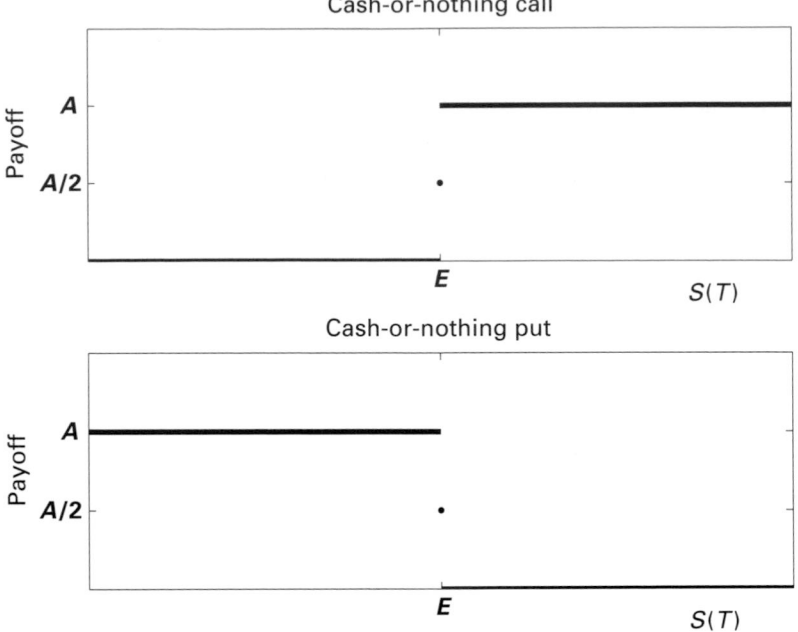

Fig. 17.1.   Payoff diagrams for cash-or-nothing call and put.

not important. But to be consistent with the formula that we derive, we will assume that $A/2$ is paid off in this at-the-money scenario, $S(T) = E$.

Analogously, a *cash-or-nothing put option* differs from a European put option in that the payoff at expiry is

$$\begin{array}{ll} 0, & \text{if } S(T) > E, \text{ and} \\ A, & \text{if } S(T) < E. \end{array}$$

Holding this option amounts to making a straight bet that the European put will finish in-the-money. As for the call, we assume that $A/2$ is paid off if $S(T) = E$. Cash-or-nothing call and put payoff diagrams are shown in Figure 17.1.

Cash-or-nothing options are sometimes called *binary*, or *digital* options, although these phrases are also used more generally when there is a discontinuous payoff diagram.

17.3 Black–Scholes for cash-or-nothing options

We will let $C^{\text{cash}}(S, t)$ and $P^{\text{cash}}(S, t)$ denote the values of the cash-or-nothing call and put options, respectively, for asset price S and time t.

The hedging argument used in Chapter 8 is very general – it requires only that the option value is a smooth function of S and t. Hence, we may ask for $C^{\text{cash}}(S, t)$

and $P^{cash}(S, t)$ to satisfy the Black–Scholes PDE (8.15). Specifying appropriate final time and boundary conditions is then sufficient to characterize the valuation formulas.

There is a simple put–call parity relation connecting $C^{cash}(S, t)$ and $P^{cash}(S, t)$, see Exercise 17.1, and hence we will focus on finding a formula for $C^{cash}(S, t)$. The cash-or-nothing call payoff function gives final time conditions

$$\lim_{t \to T^-} C^{cash}(S, t) = \begin{cases} A, & \text{for } S > E, \\ A/2, & \text{for } S = E, \\ 0, & \text{for } S < E. \end{cases} \qquad (17.1)$$

When $S = 0$, the asset remains at zero for all later times and hence the payoff is zero. This gives the boundary condition

$$C^{cash}(0, t) = 0, \qquad \text{for all } 0 \le t \le T. \qquad (17.2)$$

When S is very large, the option is almost certain to pay off the amount A. So, after discounting for interest, we find that

$$C^{cash}(S, t) \approx Ae^{-r(T-t)}, \qquad \text{for large } S. \qquad (17.3)$$

Just as for the European case, imposing the final time and boundary conditions is enough to specify a unique solution. The solution turns out to have the simple form

$$C^{cash}(S, t) = Ae^{-r(T-t)}N(d_2), \qquad (17.4)$$

where d_2 is the quantity (8.21) that appears in the European formulas. Our approach for confirming that (17.4) is an appropriate solution will be to check that the formula satisfies the Black–Scholes PDE and the extra conditions (17.1)–(17.3). Exercise 17.2 asks you to do the latter.

It is a straightforward exercise in differentiation to show that the partial derivatives appearing in the Black–Scholes PDE have the following form:

$$\frac{\partial C^{cash}}{\partial S} = \frac{Ae^{-r(T-t)}N'(d_2)}{\sigma S\sqrt{T-t}} \qquad \text{(delta)}; \qquad (17.5)$$

$$\frac{\partial^2 C^{cash}}{\partial S^2} = -\frac{Ae^{-r(T-t)}d_1 N'(d_2)}{\sigma^2 S^2(T-t)} \qquad \text{(gamma)}; \qquad (17.6)$$

$$\frac{\partial C^{cash}}{\partial t} = Are^{-r(T-t)}N(d_2)$$

$$+ Ae^{-r(T-t)}N'(d_2)\left(\frac{d_1}{2(T-t)} - \frac{r}{\sigma\sqrt{T-t}}\right) \qquad \text{(theta)}; \quad (17.7)$$

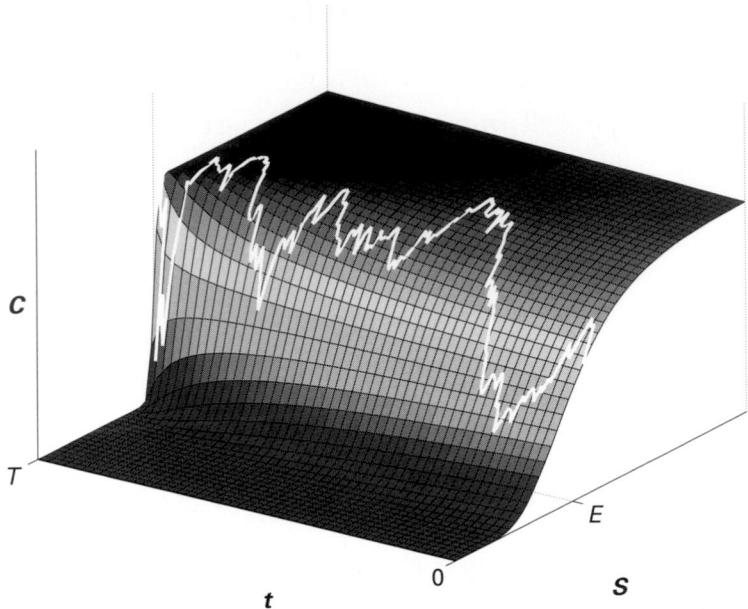

Fig. 17.2. Black–Scholes surface for a cash-or-nothing call, with asset path superimposed.

see Exercise 17.3. Inserting these expressions into the Black–Scholes PDE (8.15) we find that the expression

$$\frac{\partial C^{\text{cash}}}{\partial t} + \tfrac{1}{2}\sigma^2 S^2 \frac{\partial^2 C^{\text{cash}}}{\partial S^2} + rS\frac{\partial C^{\text{cash}}}{\partial S} - rC^{\text{cash}}$$

takes the form

$$Are^{-r(T-t)}N(d_2) + Ae^{-r(T-t)}N'(d_2)\left(\frac{d_1}{2(T-t)} - \frac{r}{\sigma\sqrt{T-t}}\right)$$
$$-\frac{1}{2}\frac{Ae^{-r(T-t)}d_1 N'(d_2)}{(T-t)} + \frac{Are^{-r(T-t)}N'(d_2)}{\sigma\sqrt{T-t}} - Are^{-r(T-t)}N(d_2),$$

which cancels down to zero, as required.

Figure 17.2 gives a plot of the surface $C^{\text{cash}}(S, t)$ and shows the option values mapped out by an asset path, in the style of Figure 11.3. For that path, the asset is close to being at-the-money near expiry, and we see that the value changes dramatically as $S(t)$ crosses the strike price E.

17.4 Delta behaviour

The delta of an option is of special interest as it plays a key role in our hedging strategy. We see from (17.5) that the call delta is always positive. This behaviour was also observed for European call options, and it can be explained in the same way – a positive payoff becomes more likely if the asset price increases.

The behaviour of the delta at expiry can be summarized as follows:

$$\lim_{t \to T^-} \frac{\partial C^{\text{cash}}}{\partial S} = \begin{cases} 0, & \text{for } S > E, \\ \infty, & \text{for } S = E, \\ 0, & \text{for } S < E, \end{cases} \tag{17.8}$$

see Exercise 17.5. Recall that the delta is precisely the amount of asset that we hold in our replicating portfolio. For the in-the-money case, $S > E$, as the time to expiry shrinks the payoff is increasingly certain to be the constant A. Since there is no risk to eliminate, we should be holding a zero amount of asset. Similarly, if the option is out-of-the-money, $S < E$, as we approach expiry, the payoff is increasingly certain to be the constant 0, which is also riskless. The infinite at-the-money delta can be thought of as a consequence of the impossibility of hedging at a point where the payoff is discontinuous. Although expiring precisely at-the-money is a probability-zero event, the delta will be large if $S(t) \approx E$ as expiry approaches. The practical consequences of a large delta are, of course, quite serious. For example, a large amount of cash needs to be withdrawn to maintain the delta-hedged portfolio and, ultimately, it will be impossible to purchase the necessary amount of asset – there will only be a finite supply available. This underlines the gap between theory and practice.

We also note that the delta behaviour summarized in (17.8) is consistent with Figure 17.2. As we approach expiry the surface starts to look like two flat, horizontal sheets joined by a vertical sheet.

In Figure 17.3 we plot the delta surface, as defined in (17.5). We chopped off the large heights that arise around $S = E$ near expiry. A path is superimposed. To emphasize that large deltas can arise, we chose an asset that stumbles towards the strike price E near expiry. The 'near infinite' deltas close to expiry are too much for the plotter to handle.

17.5 Risk neutrality for cash-or-nothing options

We saw in Chapter 12 that there is a way to derive the Black–Scholes value for a European-style option that does not make direct use of the no arbitrage principle or the concept of hedging. Instead we impose the risk neutrality assumption $\mu = r$ and compute the expected payoff, appropriately discounted for interest.

We confirm directly in this section that the idea works for a cash-or-nothing call option.

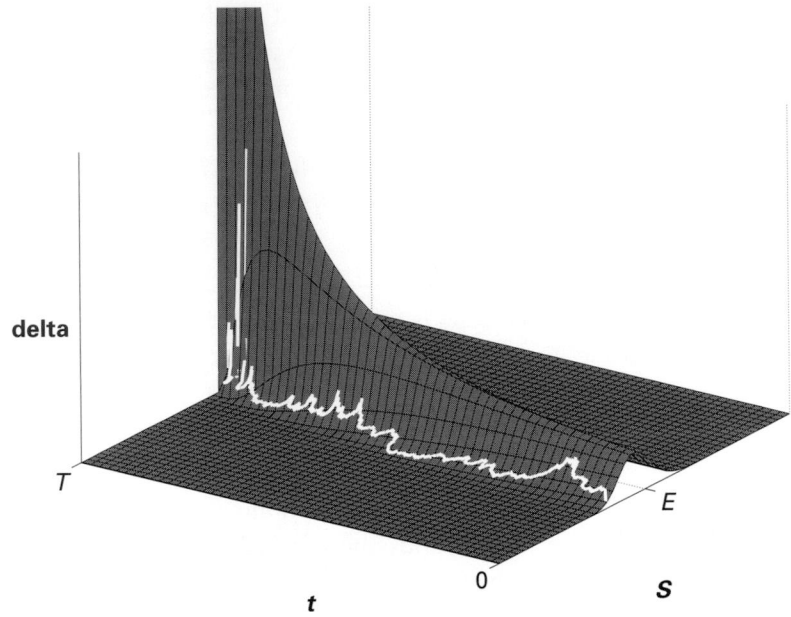

Fig. 17.3. Black–Scholes delta surface for a cash-or-nothing call, with asset path superimposed.

The payoff function $\Lambda(\cdot)$ appearing in (12.4) now has the form

$$\Lambda(x) = \begin{cases} A, & \text{for } x > E, \\ A/2, & \text{for } x = E, \\ 0, & \text{for } x < E. \end{cases}$$

Since the value of an integral does not change if we alter the value of the integrand at a single point, we may redefine $\Lambda(E) = A$, so that

$$e^{-r(T-t)}\mathbb{E}\left(\text{payoff from } S, t\right) = Ae^{-r(T-t)}\int_E^\infty \frac{\exp\left(\dfrac{-(\log(x/S)-(\mu-\frac{1}{2}\sigma^2)(T-t))^2}{2\sigma^2(T-t)}\right)}{\sigma x\sqrt{2\pi(T-t)}}\,dx. \tag{17.9}$$

Exercise 17.6 asks you to confirm that when $\mu = r$, this reduces to the Black–Scholes value $Ae^{-r(T-t)}N(d_2)$ from (17.4).

17.6 Notes and references

The terms *binary* and *digital* are not used with complete consistency in the literature. We have fixed on the unambiguous *cash-or-nothing* (and *asset-or-nothing* in Exercises 17.7 and 17.8) in line with (Hull, 2000; Nielsen, 1999).

EXERCISES

17.1. ⋆⋆⋆ By considering a portfolio consisting of a cash-or-nothing put and a cash-or-nothing call with the same strike prices and expiry dates, derive the 'cash-or-nothing put–call parity' relation

$$C^{\text{cash}}(S, t) + P^{\text{cash}}(S, t) = Ae^{-r(T-t)}. \tag{17.10}$$

17.2. ⋆⋆ Show that $C^{\text{cash}}(S, t)$ in (17.4) satisfies (17.1), (17.2) and (17.3).

17.3. ⋆⋆⋆ Differentiate (17.4) to establish (17.5), (17.6) and (17.7).

17.4. ⋆⋆⋆ Using (17.4) and (17.10), show that the value of the cash-or-nothing put option is

$$P^{\text{cash}}(S, t) = Ae^{-r(T-t)} \left(1 - N(d_2)\right).$$

Confirm that this formula has the required behaviour at $S = 0$ and in the limits $t \to T^-$ and $S \to \infty$. Also, show that it solves the Black–Scholes PDE (8.15).

17.5. ⋆⋆ Establish (17.8). (You may use without proof the fact that $\varepsilon e^{1/\varepsilon} \to \infty$ as $\varepsilon \to 0$.)

17.6. ⋆⋆⋆ Use the change of variable

$$y = -\left[\frac{\log(S/x) + (r - \frac{1}{2}\sigma^2)(T - t)}{\sigma\sqrt{T - t}}\right]$$

to show that, with $\mu = r$, the integral in (17.9) takes the value $Ae^{-r(T-t)}N(d_2)$, where d_2 is defined in (8.21).

17.7. ⋆⋆⋆ An *asset-or-nothing* call option has payoff function $\Lambda(S(T))$ of the form

$$\Lambda(x) = \begin{cases} x, & \text{for } x > E, \\ x/2, & \text{for } x = E, \\ 0, & \text{for } x < E. \end{cases}$$

Draw the payoff diagram. Show that the risk-neutral approach of setting $\mu = r$ in $e^{-r(T-t)}\mathbb{E}$ (payoff from S, t) produces the value $SN(d_1)$ for this option, where d_1 is defined in (8.20). How would the analogous asset-or-nothing put option be defined, and what is its value?

17.8. ⋆⋆ Show that holding a European call option is equivalent to holding an asset-or-nothing call option (see Exercise 17.7 above) and writing a cash-or-nothing call with $A = E$, for the same expiry date. Use this to give another way to value the asset-or-nothing call option in Exercise 17.7.

```
%CH17     Program for Chapter 17
%
% Draws Black-Scholes surface for cash-or-nothing call

clf

%%%%%%%%% Parameters %%%%%%%%%%%%%
E = 1;   A = 2;   r = 0.05;   sigma = 0.2;   T = 1;   L = 50;
%%%%%%%%%%%%%%%%%%%%%%%%%%%%%%%%%%%

Svals = linspace(0,2,L);
tvals = linspace(0,T,L);
C = zeros(L,L);

for i = 1:L
   S = Svals(i);
   for j = 1:L-1
      t = tvals(j);
      tau = T-tvals(j);
      d2 = (log(S/E) + (r - 0.5*sigma^2)*(tau))/(sigma*sqrt(tau));
      N2 = 0.5*(1+erf(d2/sqrt(2)));
      C(i,j) = A*exp(-r*tau)*N2;
   end
   % value at expiry
   C(i,L) = 0.5*A*(1+sign(S-E));
end

[Smat,tmat] = meshgrid(Svals,tvals);
mesh(Smat,tmat,C)
xlabel('t'), ylabel('S'), zlabel('C(S,t)')
```

Fig. 17.4.   Program of Chapter 17: `ch17.m`.

17.7 Program of Chapter 17 and walkthrough

In `ch17`, listed in Figure 17.4, we plot a Black–Scholes surface for a cash-or-nothing call, in the style of Figure 17.2. The code is similar to `ch11`, except that we implement the formula (17.4) directly, rather than calling a separate function. To avoid *division by zero* errors, we deal with expiry, that is, $j = L$, after the inner loop.

PROGRAMMING EXERCISES

P17.1. Alter the binomial method program `ch16` to handle the case of an asset-or-nothing call.

P17.2. Alter the discrete hedging program `ch09` to illustrate the difficulties that can arise when a cash-or-nothing option is delta-hedged.

Quotes

Markets go up; markets go down.
There is nothing insightful or sage about that observation.
Derivatives, like any other market positions, are subject to this *market risk*.
But while a normal investment may glide along a geometric path
in response to changing market conditions,
derivatives have special features that create
erratic behaviour or that accelerate or exaggerate the results.

PHILIP MCBRIDE JOHNSON (Johnson, 1999)

There are many ways to play the game,
but really only two kinds of players.
Speculators and hedgers.
Those scalping profits from churning markets,
and those seeking shelter from the storm.
Ninety-seven per cent of the daily churn in the financial markets is generated by
speculators.
This is the official figure, published by the Merc,
which defends speculators as necessary for keeping the markets
'deep' and 'liquid'.
Speculators stir the pits and hedgers pump them with funds,
and between the two of them one gets the feeding frenzy known as
the world financial markets.

THOMAS A. BASS (Bass, 1999)

18

American options

18.1 Motivation

We now look at American options. These are typically more common than Europeans. The significant new feature here is the early-exercise facility. For put options, this complicates the Black–Scholes analysis, places analytic formulas out of reach, and puts a strain on computational methods.

18.2 American call and put

An American option is like a European option except that the holder may exercise at any time between the start date and the expiry date.

Definition An *American call option* gives its *holder* the right (but not the obligation) to purchase from the *writer* a prescribed asset for a prescribed price at any time between the start date and a prescribed expiry date in the future. ◇

Definition An *American put option* gives its *holder* the right (but not the obligation) to sell to the *writer* a prescribed asset for a prescribed price at any time between the start date and a prescribed expiry date in the future. ◇

The holder of an American option is thus faced with the dilemma of deciding when, if at all, to exercise. If, at time t, the option is out-of-the-money then it is clearly best not to exercise. However, if the option is in-the-money it may be beneficial to wait until a later time where the payoff *might* be even bigger.

American options are more widely traded than their European counterparts. In many exchanges, the early-exercise feature is offered by default. It is thus important to know how much extra value, if any, this flexibility builds in.

The following argument, which is similar to one used in Section 2.6, shows that it is *never* optimal to exercise an American call option before the expiry date.

As usual, let $S(t)$ denote the asset price at time t and let E denote the exercise price. Suppose the holder wishes to exercise the option at some time $t < T$. This is only worthwhile if $S(t) > E$, and it gives a payoff of $S(t) - E$ at time t. Instead, the holder could sell the asset short at the market price at time t and then purchase the asset at time $t = T$ by doing the most favourable of

(a) exercising the option at $t = T$, and
(b) buying at the market price at time T.

With this strategy the holder has gained amount $S(t) > E$ at time t and paid out an amount less than or equal to E at time T. This is clearly better than gaining $S(t) - E$ at time t.

Since it is never optimal to exercise an American call option before the expiry date, an American call option must have the same value as a European call option. Exercise 18.1 asks you to reach this conclusion by an alternative route.

As we will see shortly, the same is not true for put options.

18.3 Black–Scholes for American options

Our aim in this section is to show how the arguments in Chapter 8 that led to the Black–Scholes PDE can be adapted to cover an American put option. We write $P^{\mathrm{Am}}(S, t)$ to denote the American put option value at asset price S and time t, and use $\Lambda(S(t)) = \max(E - S(t), 0)$ for the corresponding payoff function.

Our first observation is that

$$P^{\mathrm{Am}}(S, t) \geq \Lambda(S(t)), \qquad \text{for all } 0 \leq t \leq T, \ S \geq 0. \tag{18.1}$$

This follows from a simple arbitrage argument. If $P^{\mathrm{Am}}(S, t) < \Lambda(S(t))$ then an instantaneous profit can be made by purchasing the option and immediately exercising it. We know from Figure 11.1 that this inequality does not hold universally for a European put, so, for put options, the early-exercise feature does make a difference to the value.

Now we return to the replicating portfolio idea of Chapter 8. We may repeat the arguments up to the point (8.13) where $\Delta(V - \Pi)$, or in our case, $\Delta(P^{\mathrm{Am}} - \Pi)$, is deemed to be riskless. We now try to take the next step, which gave the equality (8.14).

Case 1: $\Delta(P^{\mathrm{Am}} - \Pi) > r(P^{\mathrm{Am}} - \Pi)$. Here, the combination $P^{\mathrm{Am}} - \Pi$ does better than cash in the bank. We argued that this could be exploited by buying $P^{\mathrm{Am}} - \Pi$, that is, buying the option and selling Π (short selling the asset and loaning out the cash).

Case 2: $\Delta(P^{\mathrm{Am}} - \Pi) < r(P^{\mathrm{Am}} - \Pi)$. Here, the combination $P^{\mathrm{Am}} - \Pi$ does worse than cash in the bank. We argued that this could be exploited by selling $P^{\mathrm{Am}} - \Pi$, that is, selling the option and buying Π (buying the asset and borrowing the cash).

Without the early exercise facility, the no arbitrage principle rules out both cases. With early exercise, however, the story changes. In Case 1, the arbitrageur buys the option and hence *controls the exercise facility*. This extra freedom can only help the arbitrageur and hence the arbitrage possibility persists. On the other hand, in Case 2 the putative arbitrageur sells the option, and is at the mercy of the early exercise facility. The arbitrageur may be exercised against at any time, and can no longer guarantee to beat the bank risklessly.

Overall, for an American put, the no arbitrage principle rules out Case 1, but not Case 2, and we conclude that (8.15) changes to

$$\frac{\partial P^{\mathrm{Am}}}{\partial t} + \tfrac{1}{2}\sigma^2 S^2 \frac{\partial^2 P^{\mathrm{Am}}}{\partial S^2} + rS \frac{\partial P^{\mathrm{Am}}}{\partial S} - r P^{\mathrm{Am}} \leq 0. \tag{18.2}$$

Note that (18.2) is a *partial differential inequality*. Now, at any point (S, t) it will be optimal to either (a) exercise, or (b) hold on to the option, and hence

$$\text{for each } S, t \text{ one of (18.1) and (18.2) is at equality.} \tag{18.3}$$

The three components (18.1), (18.2) and (18.3) are the key features in the theory of American option valuation. Together they form what is known as a *linear complementarity problem*.

At expiry, if the option is still held, its payoff matches the European, so we have the final time condition

$$P^{\mathrm{Am}}(S, T) = \Lambda(S(T)), \qquad \text{for all } S \geq 0. \tag{18.4}$$

For $S = 0$, the asset always has price zero, so a payoff of E is assured. In this case it is optimal to exercise immediately. We may interpret this formally as a boundary condition of the form

$$P^{\mathrm{Am}}(S, t) \to E, \qquad \text{as } S \to 0, \qquad \text{for all } 0 \leq t \leq T. \tag{18.5}$$

Similarly, if S is large, then the option is extremely unlikely to produce a positive payoff, so we have

$$P^{\mathrm{Am}}(S, t) \to 0, \qquad \text{as } S \to \infty, \qquad \text{for all } 0 \leq t \leq T. \tag{18.6}$$

The mathematical problem defined by (18.1)–(18.6) is much more difficult than the Black–Scholes PDE that arose without the early exercise facility. In general, there is no closed form expression for $P^{Am}(S, t)$ and we must use numerical methods to obtain approximate values.

18.4 Binomial method for an American put

It turns out that a straightforward adaptation of the binomial method can be used to value an American put. We recall from Chapter 16 that asset prices in the binomial model are determined by (16.1). If the put option is held until its expiry date then (16.2) applies. Now, working backwards through the tree, if the option is retained at time $t = t_i$ and asset price S_n^i, then the value V_n^i is given by (16.3). However, exercising the option would produce $\Lambda(S_n^i)$. Hence, choosing the best of the two possibilities leads to the relation

$$V_n^i = \max\left[\Lambda(S_n^i),\, e^{-r\delta t}\left(pV_{n+1}^{i+1} + (1-p)V_n^{i+1}\right)\right],$$

$$0 \leq n \leq i, \quad 0 \leq i \leq M - 1. \tag{18.7}$$

All together, (16.1), (16.2) and (18.7) completely specify the binomial method for computing the time-zero option value V_0^0.

Computational example We now use the binomial method to value an American put with the same parameter values as those in Section 16.4, that is, $S_0 = 9, E = 10, T = 3, r = 0.06$ and $\sigma = 0.3$. Table 18.1 shows the results for $M = 100, 200, 400$ and 1000. If we regard the $M = 1000$ result as accurate then we see that, as in the European case (Table 16.1), the method appears to converge, but does so in a nonmonotonic manner. Figures 18.1 and 18.2 give the American versions of the binomial method computations displayed in Figures 16.2 and 16.3. We see that a very similar convergence behaviour arises. Indeed, it can be shown that an error bound of the form (16.8) continues to hold.   ◇

Table 18.1. *American put value*
approximations from binomial method

	Option value
$M = 100$	1.7974
$M = 200$	1.7983
$M = 400$	1.7962
$M = 1000$	1.7962

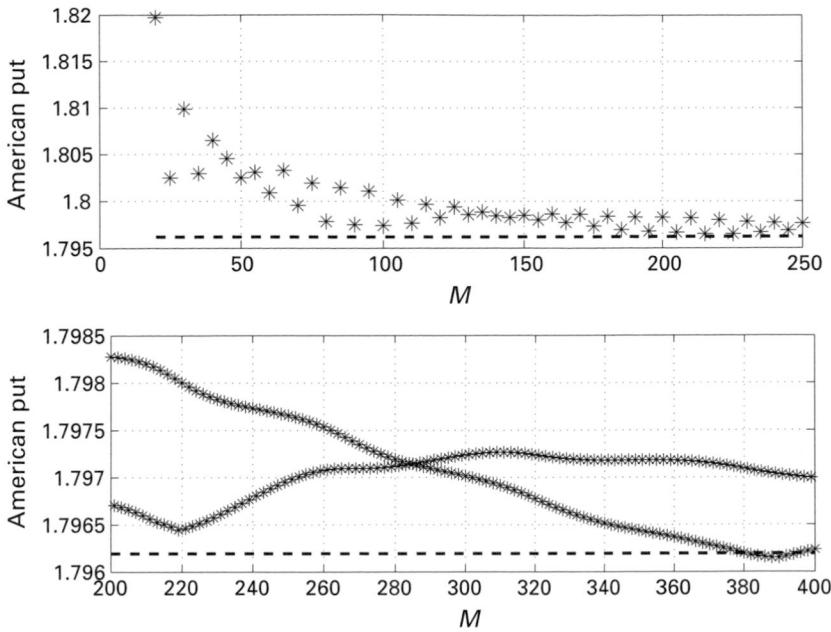

Fig. 18.1. Convergence of the binomial method for an American put as the number of time points, M, increases. Upper picture: M from 20 to 250 in steps of 5. Dashed line is 'exact' solution. Lower picture: M from 200 to 400 in steps of 1.

18.5 Optimal exercise boundary

If S is large, since there would be no payoff, it cannot be worthwhile to exercise an American put; it is optimal to hold on to the option. On the other hand, in the limit $S \to 0$, the payoff from exercising approaches the maximum possible value that we can attain; it is optimal to exercise. Interpolating between these two extremes, we might expect there to be a well-defined *optimal exercise boundary*, $S^\star(t)$, such that

- for $S(t) < S^\star(t)$ it is optimal to exercise, so $P^{\mathrm{Am}}(S, t) = \Lambda(S(t))$, and
- for $S(t) > S^\star(t)$ it is optimal to hold, so $P^{\mathrm{Am}}(S, t) > \Lambda(S(t))$.

Figure 18.3 shows the value $P^{\mathrm{Am}}(S, t)$ as a function of S, for t fixed. We set $E = 10$, $r = 0.06$, $\sigma = 0.3$ and $T = 1$, and considered $t = T/4$. We used the binomial method with a wide range of initial asset prices S_0 to compute values of $P^{\mathrm{Am}}(S, T/4)$. The figure shows that for small S the option value lies on the hockey stick $\Lambda(S(t))$, which is superimposed as a dashed line. For S bigger than some level $S^\star(T/4)$, the value $P^{\mathrm{Am}}(S, T/4)$ lies above the hockey stick. It can also be shown that the derivative $\partial P^{\mathrm{Am}}(S^\star(t), t)/\partial S = -1$, so at the point $S^\star(t)$ the curve $P^{\mathrm{Am}}(S(t), t)$ leaves the hockey stick smoothly, with a matching first derivative.

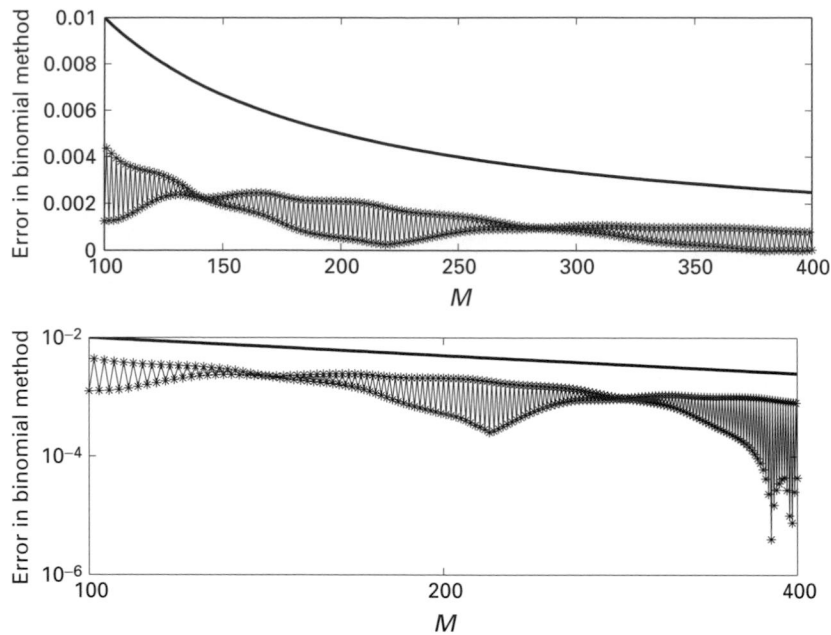

Fig. 18.2.   Upper picture: Error in the binomial method for an American put as the number of time points, M, increases from 100 to 400. Solid line is $1/M$. Lower picture: same data on a log–log scale.

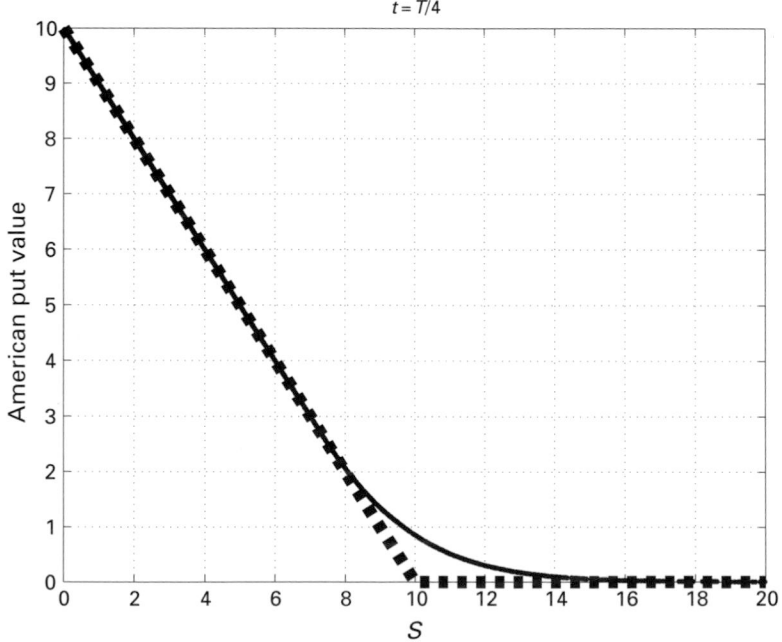

Fig. 18.3.   Value $P^{\mathrm{Am}}(S, T/4)$ for an American put, computed via the binomial method. Hockey-stick payoff function $\Lambda(S)$ is superimposed as a dashed line.

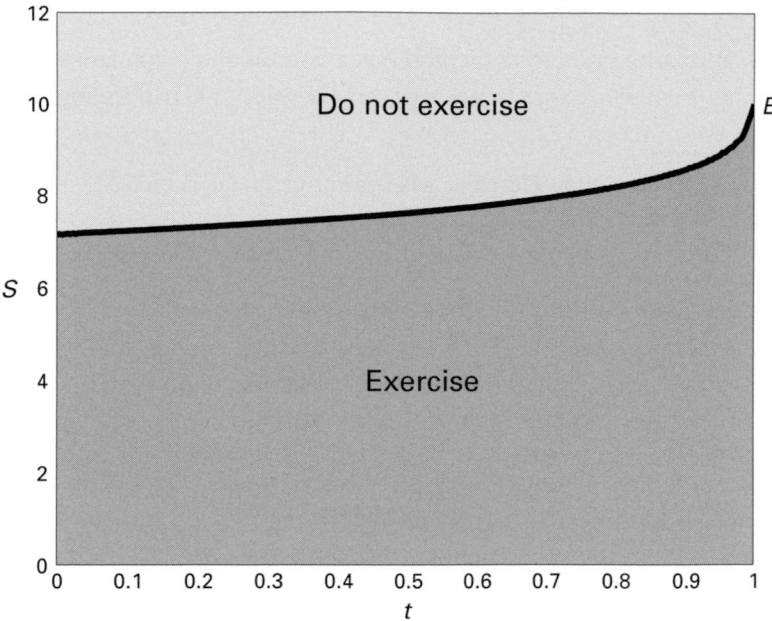

Fig. 18.4. Exercise boundary for an American put. Computed via the binomial method.

Exercise 18.2 asks you to go half-way towards proving this, by establishing -1 as a lower bound.

In Figure 18.4 we explicitly compute the optimal exercise boundary $S^\star(t)$ for the same E, r, σ and T as used in Figure 18.3. The boundary is shown as a solid curve – below this curve it is optimal to exercise and above this curve it is optimal to hold on. At $t = T/4$ we have $S^\star(t) = 7.3$, which agrees with the point on the horizontal axis in Figure 18.3 where $P^{\mathrm{Am}}(S, T/4)$ leaves the hockey stick. We tracked the optimal exercise boundary by applying the binomial method with a range of initial asset prices, S_0. At each time point, t_i, we defined $S^\star(t_i)$ to be the smallest value of S_n^i over all binomial trees for which the $e^{-r\delta t}(pV_{n+1}^{i+1} + (1-p)V_n^{i+1})$ term in (18.7) dominated the $\Lambda(S_n^i)$ term. In other words, $S^\star(t_i)$ was taken to be the smallest S_n^i for which the binomial method chose not to exercise.

It can be shown that Figure 18.4 is generic in the sense that

(i) $S^\star(T) = E$,
(ii) $S^\star(t)$ is a well-defined, single-valued function of t,
(iii) $S^\star(t)$ is a nondecreasing function of t.

Exercise 18.3 deals with points (i) and (iii).

18.6 Monte Carlo for an American put

We have seen that the binomial method has a natural extension from European to American options. The same is not true for the Monte Carlo method. This mismatch has two sources.

(a) Monte Carlo deals with single paths, whereas the binomial method essentially averages over paths automatically.
(b) Monte Carlo works forward in time, whereas the binomial method runs backwards.

Monte Carlo for European options exploits the idea that the value can be expressed as an expectation. In the American case there is an analogous, but less computationally useful, representation. Under the risk neutrality condition $\mu = r$, the time-zero American put value may be expressed as

$$P^{\text{Am}}(S_0, 0) = \sup_{0 \le \tau \le T} \mathbb{E}\left[e^{-r\tau} \Lambda(S(\tau))\right], \tag{18.8}$$

where τ is a *stopping time*. To define a stopping time precisely requires technicalities that have not been developed in this book, but the expression (18.8) can be described informally as follows.

- The value taken by τ determines the time at which the option is exercised. So $e^{-r\tau} \Lambda(S(\tau))$ in (18.8) represents the discounted payoff.
- The quantity τ is a random variable that depends upon the asset path $S(t)$.
- Any rule that specifies τ as a function of the asset path $S(t)$ can be used, with the proviso that the decision to set $\tau = t^*$ can only use information about $S(t)$ for $0 \le t \le t^*$.
- The option value $P^{\text{Am}}(S_0, 0)$ is given by using the rule for determining τ that leads to the biggest expected payoff, suitably discounted for interest.

Putting this in words:

Imagine all possible exercise strategies, that is, all possible rules for determining when to exercise the option. Suppose we judge the success of a strategy by its discounted expected payoff. Then we recover the Black–Scholes American put option value if we use the best out of all those exercise strategies that do not look forward in time – those that take an exercise decision at each point in time using only information about the asset price up to that time.

From a computational perspective, an enormous hurdle in (18.8) is the requirement to optimize over all allowable exercise strategies. It is impossible to write down all such strategies in any useful way, let alone optimize over them! To illustrate the idea, we restrict ourselves to a very simple class of allowable exercise strategies. Suppose we decide to exercise the option at time t if the discounted payoff, $e^{-rt} \Lambda(S(t))$, exceeds some fixed level $\alpha > 0$. If we reach the expiry date, T, and have not yet exercised the option, then it makes sense to exercise if $\Lambda(S(T)) > 0$. Overall, our exercise strategy may be written as follows.

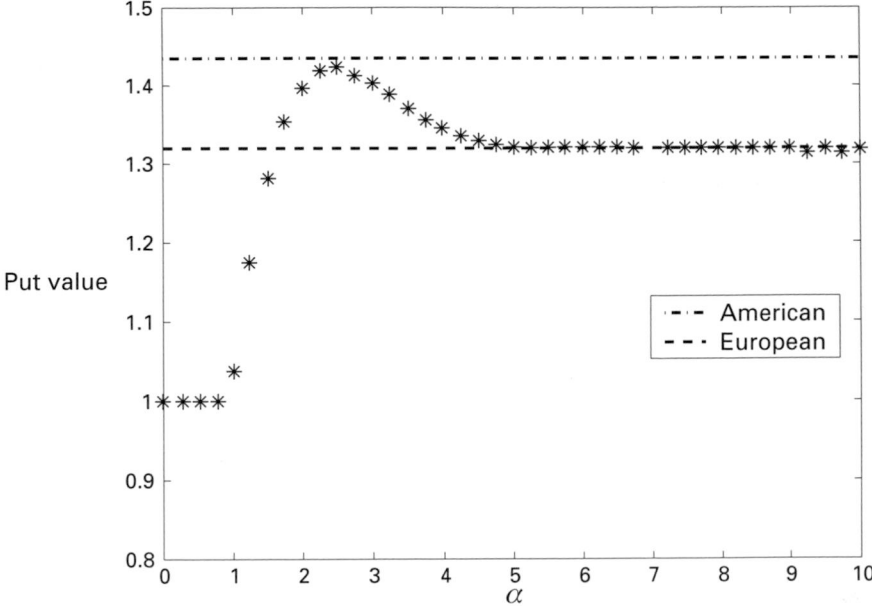

Fig. 18.5. Asterisks are Monte Carlo approximations to the discounted expected American put payoff with a simple exercise strategy parametrized by α. Upper and lower horizontal lines show the true American and European values.

- Exercise at time t if $e^{-rt}[E - S(t)] > \alpha$.
- If we reach T, exercise if $E - S(T) > 0$.

This is an allowable strategy, as the decision about whether to exercise at time t uses only $S(t)$. In Figure 18.5 we measure the success of this approach. Here we valued an American option with $S_0 = 9$, $E = 10$, $T = 1$, $r = 0.06$ and $\sigma = 0.3$. The Black–Scholes value, computed via the binomial method, was found to be 1.43. The corresponding European put option value is 1.32. These values are indicated as horizontal lines. The asterisks in the figure show the Monte Carlo approximations to the option value, using the exercise strategy above, with a range of choices for α. More precisely, we computed 10^6 risk-neutral discrete asset paths, with a time spacing of $\delta t = 10^{-3}$, and applied the strategy at each discrete time point $i\delta t$. Confidence intervals for the sample means were smaller than the size of the asterisks in the plot. We see from the figure that if α is taken to be around 2.5, the discounted expected payoff is close to the Black–Scholes value. Exercise 18.4 asks you to explain the results for $0 \le \alpha \le 1$ and α large. In this example, we are fortunate that optimizing over the parameter α in our simple class of exercise strategies gives an answer that is close to the optimal over all allowable strategies. Of course, if we were to change S_0, E, T, r or σ then the optimal α would certainly change, and there is no guarantee that it would give a good approximation to $P^{Am}(S_0, 0)$.

In general, picking any particular allowable strategy and computing the discounted expected payoff will lead to a lower bound on the true Black–Scholes value.

By contrast, we could allow ourselves the luxury of peeking into the future in order to select the best possible exercise times.

- Consider the whole path $S(t)$ for $0 \leq t \leq T$, and exercise where $e^{-rt} \Lambda(S(t))$ is maximized.

For each asset path, this strategy does at least as well as the best allowable strategy. Hence, the corresponding discounted expected payoff gives an upper bound on the Black–Scholes value. In the example of Figure 18.5 the upper bound was 2.62, which, as is typical, is too crude to be of much use.

18.7 Notes and references

Our derivation of the linear complementarity problem (18.2)–(18.6) followed closely the treatment by Almgren (Almgren, 2002). It is possible to write the American put valuation problem in terms of a PDE that explicitly involves the optimal exercise boundary, $S^\star(t)$. This *free boundary problem* approach is described in (Kwok, 1998; Wilmott *et al.*, 1995), for example. Kwok (Kwok, 1998) gives examples of more complex options with early exercise features for which the exercise and non-exercise regions are made up of disconnected sets. The condition that $\partial P^{Am}(S^\star(t), t)/\partial S = -1$, which we illustrated in Figure 18.3, is discussed in detail in (Kwok, 1998) and (Wilmott *et al.*, 1995).

Convergence of the binomial method for American options is treated in (Leisen, 1998), where an error bound of the form (16.8) is derived.

The argument in Section 18.2 that shows the equivalence of European and American call values fails to hold when the asset pays dividends, see (Hull, 2000; Kwok, 1998; Wilmott *et al.*, 1995), for example, for details of how the theory can be adapted.

Applied mathematicians have recently become interested in the nature of the optimal exercise boundary for $t \approx T$. It can be shown that as the boundary $S^\star(t)$ approaches E as $t \rightarrow T^-$, its tangent becomes unbounded, as may be observed in Figure 18.4. The precise nature of this singularity is explored in (Goodman and Ostrov, 2002; Kuske and Keller, 1998), for example.

Björk (Björk, 1998) is a good source for the mathematics behind (18.8).

Until quite recently, most researchers believed that a Monte Carlo approach could not be used for valuing American options. However, a number of authors now argue that, with appropriate extensions, competitive Monte Carlo based computational algorithms are achievable; see (Anderson and Broadie, 2001; Boyle *et al.*, 1997; Fu *et al.*, 2001; Longstaff and Schwartz, 2001; Rogers, 2002), for example.

EXERCISES

18.1. ★★★ Repeat the analysis in Section 18.3 for the case of an American call option. Show that the Black–Scholes European call option formula (8.19) satisfies the relevant analogues of (18.2)–(18.6). Deduce that an American call option has the same value as the corresponding European call option.

18.2. ★★ In Section 18.5 it was mentioned that $\partial P^{\mathrm{Am}}(S^\star(t), t)/\partial S = -1$. Give a simple explanation why $\partial P^{\mathrm{Am}}(S^\star(t), t)/\partial S$ cannot be less that -1.

18.3. ★★★ Given that there is a well-defined, single-valued optimal exercise boundary function $S^\star(t)$ for an American put, show that $S^\star(T) = E$ and that $S^\star(t)$ is a nondecreasing function of t.

18.4. ★★ Explain the behaviour of the Monte Carlo approximations in Figure 18.5 for $0 \le \alpha \le 1$ and α large.

18.5. ★★ Which of the following exercise strategies are allowable in (18.8)?

Strategy 1:

- Exercise at time t if $S(t) < \frac{1}{2}E$.
- If we reach T, exercise if $E - S(T) > 0$.

Strategy 2:

- Exercise at time t if $S(t) < \min(E, 1.1 \min_{0 \le r \le T} S(r))$.
- If we reach T, exercise if $E - S(T) > 0$.

Strategy 3:

- Exercise at time t if $S(t) < \min(E, \frac{1}{2} \min_{0 \le r \le t/2} S(r))$.
- If we reach T, exercise if $E - S(T) > 0$.

18.8 Program of Chapter 18 and walkthrough

In ch18, listed in Figure 18.6, we give a modified version of ch16 that values an American put with the binomial method. After initializing parameters, we create the one-dimensional array dpowers with entries $d^M, d^{M-1}, \ldots, d^0$ and the one-dimensional array upowers with entries $u^0, u^1, \ldots, u^M$. It follows that S*dpowers.*upowers gives the asset values $S_0^M, S_1^M, \ldots, S_M^M$ at expiry in the asset price tree of Figure 16.1, and

```
S*dpowers(M-i+2:M+1).*upowers(1:i);
```

```
%CH18      Program for Chapter 18
%
% Implements binomial method for an American put.

%%%%% Problem and method parameters %%%%%%%%%
S = 3; E = 4; T = 1; r = 0.05; sigma = 0.3;
M = 400; dt = T/M; p =0.5;
u = exp(sigma*sqrt(dt) + (r-0.5*sigma^2)*dt);
d = exp(-sigma*sqrt(dt) + (r-0.5*sigma^2)*dt);
%%%%%%%%%%%%%%%%%%%%%%%%%%%%%%%%%%%%%%%%

% Initial computations
dpowers = d.^([M:-1:0]');
upowers = u.^([0:M]');

% Time T option values
W = max(E-S*dpowers.*upowers,0);

% Work back to option value at time zero
for i = M:-1:1
    Si = S*dpowers(M-i+2:M+1).*upowers(1:i);
    W = max(max(E-Si,0),exp(-r*dt)*(p*W(2:i+1)+(1-p)*W(1:i)));
end

disp('Option value is'), disp(W)
```

Fig. 18.6.   Program of Chapter 18: ch18.m.

gives the asset values $S_0^i, S_1^i, \ldots, S_i^i$ at the ith time level. In this way, the iteration (18.7) is encapsulated as

$$\texttt{W = max(max(E-Si,0),exp(-r*dt)*(p*W(2:i+1)+(1-p)*W(1:i)));}$$

As with ch16, the loops exits with a scalar value for W that gives the option value V_0^0.

   The option value output by ch18.m is 1.0158. The validity of the result will be confirmed by ch24 in Chapter 24.

PROGRAMMING EXERCISES

P18.1. Alter ch18 in order to re-create Figure 18.4.

P18.2. Think up an allowable exercise strategy and test it in the manner of Figure 18.5.

Quotes

Although simulation is a powerful tool
for solving some higher-dimensional problems,

conventional wisdom was that
simulation could not be applied to American-style pricing problems.
The algorithms described here represent the first attempts to solve these problems
that were long thought to be computationally intractable.

> PHELIM BOYLE, MARK BROADIE AND PAUL GLASSERMAN (Boyle *et al.*, 1997)

Academia was teeming with nerdy mathematicians who had been publishing
unintelligible dissertations on markets for years.
Wall Street had started to hire them, but only for research,
where they'd be out of harm's way.
On Wall Street, the eggheads were stigmatized as 'quants',
unfit for the man's game of trading.

> ROGER LOWENSTEIN (Lowenstein, 2001)

I prefer the judgement of a 55-year old trader
to that of a 25-year old mathematician.

> ALAN GREENSPAN, source (Taleb, 1997)

19

Exotic options

19.1 Motivation

So far, we have seen European options and American-style options. A bewildering array of alternatives are also available; these go by the general name of *exotic options*. Each type of option is distinguished by

(i) the nature of its *path dependency* – the way in which the payoff depends upon the asset path $S(t)$ for $0 \leq t \leq T$, and

(ii) whether early exercise is allowed.

In many cases, exact expressions for the option value are not available, and hence approximations must be computed. This chapter introduces some of the less esoteric exotics and discusses the use of our two computational algorithms: the binomial and Monte Carlo methods. A third computational approach, numerical solution of a Black–Scholes PDE formulation, is covered in Chapters 23 and 24.

19.2 Barrier options

Barrier options have a payoff that switches on or off depending on whether the asset crosses a pre-defined level.

- A *down-and-out call* option has a payoff that is zero **if** the asset crosses some pre-defined barrier $B < S_0$ at some time in $[0, T]$. If the barrier is not crossed then the payoff becomes that of a European call, $\max(S(T) - E, 0)$.

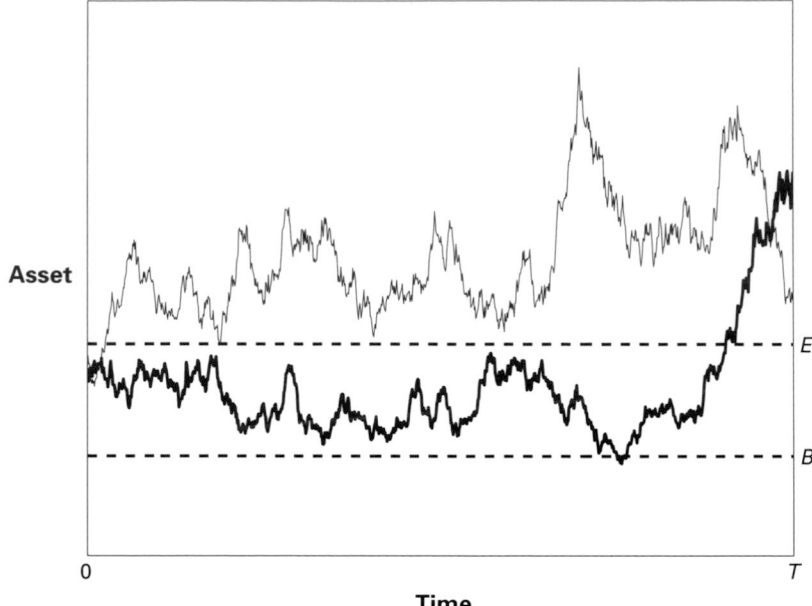

Fig. 19.1. Two asset paths and a barrier, B. The thicker asset path crosses the barrier and hence would give zero payoff in a down-and-out call. The thinner asset path fails to cross the barrier and hence would give zero payoff in a down-and-in call.

- A *down-and-in call* option has a payoff that is zero **unless** the asset crosses some pre-defined barrier $B < S_0$ at some time in $[0, T]$. If the barrier is crossed then the payoff becomes that of a European call, $\max(S(T) - E, 0)$.

One reason for the popularity of barrier options is that, because the payoff opportunities are more limited, they are cheaper to buy than Europeans. Figure 19.1 illustrates the idea. Here, two asset paths are shown. Both expire above the exercise price: $S(T) > E$. Despite finishing the higher, the thicker of the two paths dips lower, crossing the barrier. The thicker path would give a nonzero payoff for a down-and-in call, but a zero payoff for a down-and-out call. Conversely, the thinner path would give a zero payoff for a down-and-in call, but a nonzero payoff for a down-and-out call.

The hedging idea from Chapter 8 remains valid for barrier options. Let $C^{\mathrm{B}}(S, t)$ denote the value of a down-and-out call option at asset price S and time t. The Black–Scholes PDE (8.15) is relevant unless the barrier is crossed, so $C^{\mathrm{B}}(S, t)$ must satisfy the PDE on the domain $0 \leq t \leq T$, $B \leq S$. If $S = B$ then the option becomes worthless, giving

$$C^{\mathrm{B}}(B, t) = 0, \qquad \text{for } 0 \leq t \leq T. \tag{19.1}$$

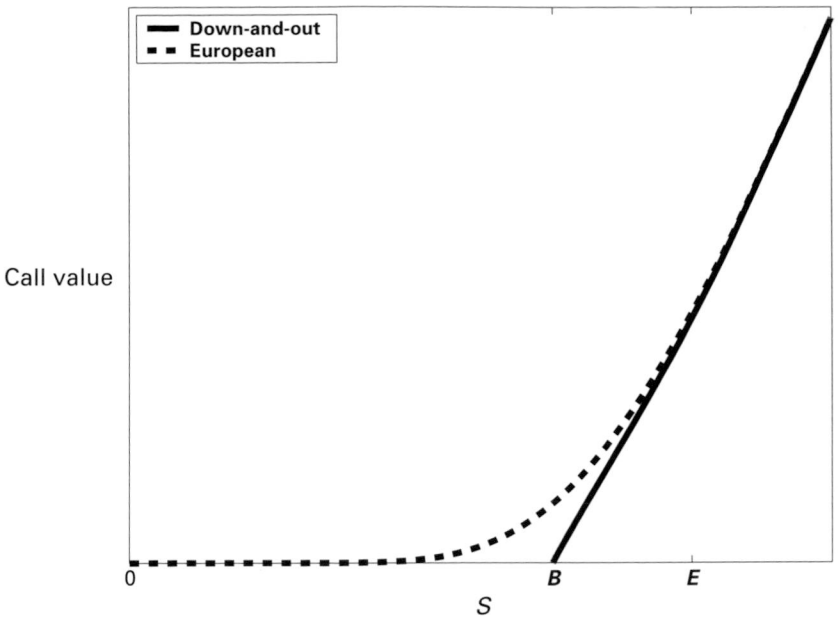

Fig. 19.2.   Time-zero down-and-out call value (19.3) as a function of S.

Also, at expiry, for $S(T) > B$ we must recover the European value, so

$$C^{\mathrm{B}}(S, T) = C(S, T), \qquad \text{for } B \le S. \tag{19.2}$$

Here, $C(S, t)$ denotes the European value (8.19). In the case $B < E$ it can be shown that a solution to the Black–Scholes PDE on the domain $0 \le t \le T, B \le S$, that satisfies (19.1) and (19.2) is given by

$$C^{\mathrm{B}}(S, t) = C(S, t) - (S/B)^{1-2r/\sigma^2} C(B^2/S, t); \tag{19.3}$$

see Exercise 19.2.

We note that (19.3) immediately confirms that the down-and-out call is worth less than the European call.

A plot of the time-zero value $C^{\mathrm{B}}(S, 0)$ in (19.3) for $B < E$ is given in Figure 19.2. The European value is also shown. As we would expect, as the initial asset price increases, and so the probability of hitting the barrier decreases, the down-and-out call value approaches that of the European.

Given a formula for a down-and-out call, the corresponding down-and-in can be found from the relation

$$\text{in } + \text{ out} = \text{European}, \tag{19.4}$$

see Exercise 19.3.

Replacing 'down' by 'up' gives another class of barrier options.

- An *up-and-out call* option has a payoff that is zero **if** the asset crosses some pre-defined barrier $B > S_0$ at some time in $[0, T]$. If the barrier is not crossed then the payoff becomes that of a European call, $\max(S(T) - E, 0)$.
- An *up-and-in call* option has a payoff that is zero **unless** the asset crosses some predefined barrier $B > S_0$ at some time in $[0, T]$. If the barrier is crossed then the payoff becomes that of a European call, $\max(S(T) - E, 0)$.

There are also, of course, put versions of the above calls; just replace the word 'call' by 'put' in each case. This gives a total of eight different up/down-and-in/out calls/puts. In each case, an analytical formula for the option value can be obtained by solving the Black–Scholes PDE with appropriate final time and boundary conditions. Formulas for each type of barrier option can be found via the references in Section 19.7. As an example that we will return to in Section 19.6, we give the formula for an up-and-out call:

$$
S \left(N(d_1) - N(e_1) - \left(\frac{B}{S} \right)^{1+2r/\sigma^2} (N(f_2) - N(g_2)) \right)
$$

$$
- E e^{-r(T-t)} \left(N(d_2) - N(e_2) - \left(\frac{B}{S} \right)^{-1+2r/\sigma^2} (N(f_1) - N(g_1)) \right). \quad (19.5)
$$

Here, d_1 and d_2 are defined in (8.20) and (8.21) and

$$
e_1 = \frac{\log(S/B) + (r + \frac{1}{2}\sigma^2)(T - t)}{\sigma \sqrt{T - t}},
$$

$$
e_2 = \frac{\log(S/B) + (r - \frac{1}{2}\sigma^2)(T - t)}{\sigma \sqrt{T - t}},
$$

$$
f_1 = \frac{\log(S/B) - (r - \frac{1}{2}\sigma^2)(T - t)}{\sigma \sqrt{T - t}},
$$

$$
f_2 = \frac{\log(S/B) - (r + \frac{1}{2}\sigma^2)(T - t)}{\sigma \sqrt{T - t}},
$$

$$
g_1 = \frac{\log(SE/B^2) - (r - \frac{1}{2}\sigma^2)(T - t)}{\sigma \sqrt{T - t}},
$$

$$
g_2 = \frac{\log(SE/B^2) - (r + \frac{1}{2}\sigma^2)(T - t)}{\sigma \sqrt{T - t}}.
$$

Figure 19.3 plots the up-and-out call value (19.5) at time zero, along with the corresponding European. The picture illustrates that barrier options can be significantly cheaper than Europeans. The up-and-out call has a limited up-side – the

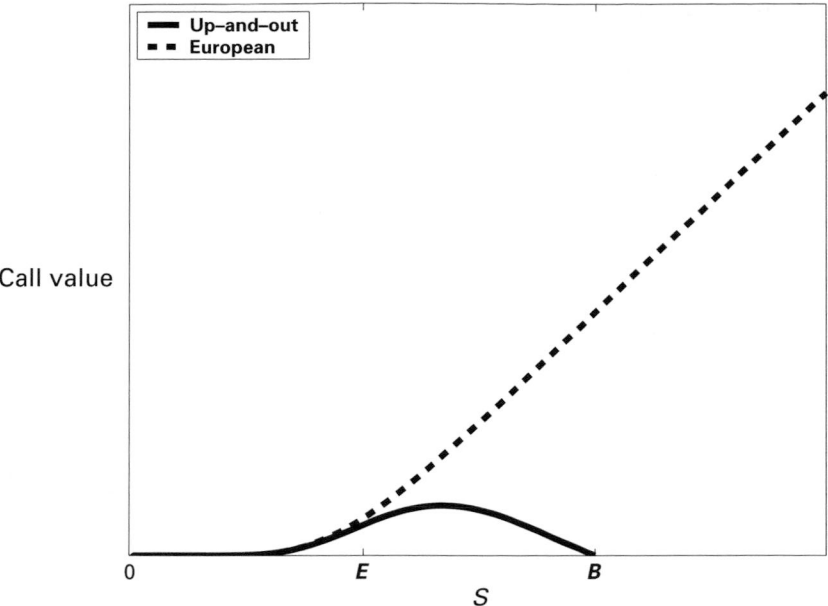

Fig. 19.3.   Time-zero up-and-out call value (19.5) as a function of S.

payoff cannot exceed $B - E$, and hence can be bought for much less than the European version.

There are many generalizations of those eight basic barrier options.

- *Double barrier* options impose upper and lower bounds on the asset price, and payoff may knock in (or out) if either barrier is (or both barriers are) crossed.
- *Partial barrier* options have barriers that apply for a limited time interval.
- *Parisian* options have barriers that must remain crossed for some pre-specified amount of time.
- More generally, the barrier may be time-dependent and the nature of the option may be re-set (e.g. to another barrier option) if a barrier is crossed.

Although the Black–Scholes analysis remains relevant in all cases, the more complicated barrier options do not admit analytical expressions for the value.

19.3 Lookback options

The payoff for a lookback option depends upon either the maximum or the minimum value attained by the asset. There are two broad categories, fixed and floating strikes. In describing them, we use the notation

$$S^{\max} := \max_{[0,T]} S(t) \qquad \text{and} \qquad S^{\min} := \min_{[0,T]} S(t)$$

to denote the extreme asset values.

- A *fixed strike lookback call* option has payoff at the expiry date T given by $\max(S^{\max} - E, 0)$.
- A *fixed strike lookback put* option has payoff at the expiry date T given by $\max(E - S^{\min}, 0)$.
- A *floating strike lookback call* option has payoff at the expiry date T given by $S(T) - S_{\min}$.
- A *floating strike lookback put* option has payoff at the expiry date T given by $S^{\max} - S(T)$.

These lookback options are clearly more valuable than the corresponding Europeans. The fixed strike lookbacks differ from European options in that the final asset value $S(T)$ is replaced by the 'best' asset price – the maximum in the case of a call and the minimum in the case of a put. With a floating strike, the exercise (strike) price becomes the extremely favourable minimum asset price for a call and maximum asset price for a put. In the floating case it will always be worthwhile to exercise, so the word 'option' is perhaps inappropriate.

It is possible to derive Black–Scholes formulas for the four lookback cases above, see Section 19.7 for references. There are many extensions of these ideas, typically designed to offer some of the lookback desirability at a cheaper price; for example by looking back over a limited time period or over a finite number of points in time. In many cases, the options may only be valued by computational means.

19.4 Asian options

Whereas barriers and lookbacks focus on extreme values of the asset, Asian options are determined by average case behaviour.

- An *average price Asian call* option has payoff at the expiry date T given by

$$\max\left(\frac{1}{T}\int_0^T S(\tau)d\tau - E, 0\right).$$

- An *average price Asian put* option has payoff at the expiry date T given by

$$\max\left(E - \frac{1}{T}\int_0^T S(\tau)d\tau, 0\right).$$

Here we are replacing the final asset price $S(T)$ that would be used in a European option by the *average* asset price over the time period.

- An *average strike Asian call* option has payoff at the expiry date T given by

$$\max\left(S(T) - \frac{1}{T}\int_0^T S(\tau)d\tau, 0\right).$$

- An *average strike Asian put* option has payoff at the expiry date T given by

$$\max \left(\frac{1}{T} \int_0^T S(\tau)d\tau - S(T), 0 \right).$$

Here we are replacing the strike, or exercise, price E, that would be used in a European option, by the *average* asset price.

Other Asian options can be defined, for instance, by replacing the *continuous average* $\int_0^T S(\tau)d\tau/T$ by an *arithmetic average*

$$\frac{1}{n} \sum_{i=1}^n S(t_i),$$

or *geometric average*

$$\left(\prod_{i=1}^n S(t_i) \right)^{1/n},$$

over n time points, $0 \le t_1 < t_2 < \cdots < t_n \le T$. (In practice, as the real asset price does not change continuously, even the continuous average would have to be approximated from discrete market data.)

The path dependency for Asians is, in a sense, more complicated than that for barrier and lookback options. The payoff depends on the range of asset prices, not just the extremes. It is possible to accommodate Asian options into the Black–Scholes framework, but exact solutions have been found only in certain cases. One such case is treated in Exercise 19.6.

19.5 Bermudan and shout options

A *Bermudan* option differs from the corresponding American option in only one respect. While the American option allows the holder to exercise at any time in $[0, T]$, the Bermudan option restricts the early exercise facility to a fixed number of pre-determined dates.

As in the American case, there is no general analytical formula for the Bermudan option value.

The simplest version of a *shout call* option allows the holder to 'shout' at most once to the writer between times 0 and T. The payoff at expiry is given by

$$\left\{ \begin{array}{ll} \max(S(T) - E, S(\tau) - E), & \text{if holder shouted at time } \tau, \\ \max(S(T) - E, 0), & \text{if holder did not shout,} \end{array} \right\} \quad (19.6)$$

and we may make the perfectly sensible assumption that a shout will only take place if $S(\tau) > E$. The effect of shouting is to lock in a payoff of at least $S(\tau) - E$; the actual payoff will then be the maximum of this value and the European payoff. Typically, a shout will take place if the holder feels that the asset price has peaked and is about to plummet. As with Americans and Bermudans, there is no exact valuation formula for shouts.

19.6 Monte Carlo and binomial for exotics

The Monte Carlo method that we described in Chapter 15 extends easily to handle path dependency. The extra step required is to set up a grid of points $t_j = j\Delta t$, for $0 \le j \le N$, where N is a large number and $\Delta t = T/N$. We are given $S(0) = S_0$, so from (6.9) we can compute an asset price $S(t_{j+1})$ in terms of $S(t_j)$ using the formula

$$S(t_{j+1}) = S(t_j)e^{(r-\frac{1}{2}\sigma^2)\Delta t + \sigma\sqrt{\Delta t}Z_j}, \qquad \text{for i.i.d. } Z_j \sim \mathsf{N}(0, 1). \qquad (19.7)$$

(Note that we use the risk neutrality assumption, $\mu = r$.) This gives us the asset price at a closely spaced set of points in $[0, T]$, so we can compute approximations to the max, min or integral, and test for barrier crossings. For example, the following algorithm values an up-and-out call option. Here, M is the number of asset paths that we sample.

> for $i = 1$ to M
>> for $j = 0$ to $N - 1$
>>> compute an $\mathsf{N}(0, 1)$ sample ξ_j
>>> set $S_{j+1} = S_j e^{(r-\frac{1}{2}\sigma^2)\Delta t + \sigma\sqrt{\Delta t}\xi_j}$
>> end
>> set $S_i^{\mathrm{max}} = \max_{0 \le j \le N} S_j$
>> if $S_i^{\mathrm{max}} < B$ set $V_i = e^{-rT}\max(S_N - E, 0)$, otherwise set $V_i = 0$
> end
> set $a_M = \frac{1}{M}\sum_{i=1}^M V_i$
> set $b_M^2 = \frac{1}{M-1}\sum_{i=1}^M (V_i - a_M)^2$

The result gives an approximate option price a_M and an approximate 95% confidence interval (15.5).

For Asian options we could use the Riemann sum $\Delta t \sum_{j=1}^N S_j$ to approximate the integral $\int_0^T S(\tau)d\tau$. With an average price Asian put this would give the following algorithm:

> for $i = 1$ to M
>> for $j = 0$ to $N - 1$
>>> compute an $\mathsf{N}(0, 1)$ sample ξ_j
>>> set $S_{j+1} = S_j e^{(r-\frac{1}{2}\sigma^2)\Delta t + \sigma\sqrt{\Delta t}\xi_j}$
>> end
>> set $Smean_i = (\Delta t/T)\sum_{j=1}^N S_j$

Table 19.1. *Ninety-five per cent confidence intervals for Monte Carlo on a European up-and-out call. Black–Scholes value (19.5) is 0.0983*

	$\Delta t = 10^{-2}$	$\Delta t = 10^{-3}$	$\Delta t = 10^{-4}$
$M = 10^2$	[0.0469, 0.1671]	[0.0397, 0.1387]	[0.0569, 0.1813]
$M = 10^3$	[0.0961, 0.1347]	[0.0756, 0.1104]	[0.0726, 0.1046]
$M = 10^4$	[0.1042, 0.1163]	[0.0997, 0.1112]	[0.0926, 0.1038]
$M = 10^5$	[0.1097, 0.1136]	[0.1000, 0.1036]	[0.0981, 0.1071]

$$\text{set } V_i = e^{-rT} \max(E - Smean_i, 0)$$
end
$$\text{set } a_M = \frac{1}{M} \sum_{i=1}^{M} V_i$$
$$\text{set } b_M^2 = \frac{1}{(M-1)} \sum_{i=1}^{M} (V_i - a_M)^2$$

Computational example We now apply Monte Carlo to the task of valuing an up-and-out call with $S_0 = 5$, $E = 6$, $\sigma = 0.3$, $r = 0.05$ and $T = 1$, with barrier $B = 8$. The Black–Scholes value (19.5) was found to be 0.0983. Table 19.1 shows the 95% confidence intervals for timesteps Δt of 10^{-2}, 10^{-3} and 10^{-4} (so $N = 10^2$, 10^3 and 10^4), and number of discrete sample paths M equal to 10^2, 10^3, 10^4 and 10^5. As the theory predicts, increasing M causes the confidence interval to shrink. However, in general the Monte Carlo method is over-estimating the option value. In particular, even for the largest sample size, $M = 10^5$, the $\Delta t = 10^{-2}$ and $\Delta t = 10^{-3}$ confidence intervals do not contain the Black–Scholes value. To understand why, recall that the method is sampling the path at finitely many discrete points, rather than over the continuous interval $[0, T]$. Because the discrete test $\max_{0 \leq j \leq N} S_j < B$ is less stringent than the continuous test $\max_{0 \leq t \leq T} S(t) < B$, the Monte Carlo method allows more nonzero payoffs than it should. As Δt is refined (so N increases) the discrete test approaches the continuous one, and the bias becomes less pronounced. In Table 19.1, we see that for $\Delta t = 10^{-4}$ and $M = 10^5$, the confidence interval does contain the Black–Scholes value, although it is still skewed to the right. A more expensive simulation with $\Delta t = 10^{-5}$ and $M = 10^6$ improved the confidence interval to [0.0980, 0.0992]. ◇

Although the Monte Carlo method typically produces low-accuracy solutions, it does have the benefit of flexibility. It should be clear that the pathwise sampling approach can be applied to any of the generalized path-dependent options mentioned in Sections 19.2, 19.3, 19.4.

The binomial method does not naturally extend to path-dependent options, as the basic recombining tree of asset prices in Figure 16.1 loses track of individual asset paths. At time $t_M = T$ we have only a set of asset prices $\{S_n^M\}_{n=0}^M$ and no information about how those asset prices were reached. (In fact we are essentially averaging over all paths that finished at that price). Even so, researchers have developed techniques for adapting the binomial method to barriers, lookbacks and Asians; see Section 19.7 for references.

Conversely, as we have seen in Chapter 18, early exercise does not fit comfortably with Monte Carlo, but is easily incorporated into the binomial method.

In the case of Bermudan options, it is clear that the binomial method may be used. In fact, as applied to American options in Section 18.4, the method is really approximating the American by a Bermudan with a large number of closely spaced early exercise points. Bermudan options can thus be handled this way if we simply make sure that the prescribed exercise dates are included in the set of times t_i, and then use (18.7) if t_i is an allowable exercise time and (16.3) otherwise.

To handle the shout option with payoff (19.6), note that if a shout happened at time τ, then the payoff may be written

$$\max\left(S(T) - S(\tau), 0\right) + S(\tau) - E. \tag{19.8}$$

From this point of view, a shout locks in a bonus of $S(\tau) - E$ and moves the exercise price to $S(\tau)$. Once τ and $S(\tau)$ are known, the first term in (19.8), $\max(S(T) - S(\tau), 0)$, corresponds to the payoff for a European option, so it is given by the Black–Scholes formula (8.19) with time set to τ and exercise price set to $S(\tau)$. We may thus use the approach outlined in Section 18.4 with (18.7) replaced by

$$V_n^i = \max\Big[\text{value (19.8) from shouting at } (t_i, S_n^i),$$
$$e^{-r\delta t}\left(pV_{n+1}^{i+1} + (1-p)V_n^{i+1}\right)\Big], \tag{19.9}$$

for $0 \leq n \leq i$ and $0 \leq i \leq M - 1$. The overall method is then defined by (16.1), (16.2) and (19.9).

19.7 Notes and references

The texts (Kwok, 1998) and (Wilmott *et al.*, 1995), and any of the Wilmott incarnations, such as (Wilmott, 1998), give much more detail about how the Black–Scholes PDE framework can be used to value exotic options. Also (Hull, 2000; Kwok, 1998; Wilmott, 1998) are good sources for analytical valuation formulas.

Chapter 13 of (Björk, 1998) deals with barriers and lookbacks from a martingale/risk-neutral perspective.

The use of the binomial method for barriers, lookbacks and Asians is discussed in (Hull, 2000; Kwok, 1998).

There are many ways in which the features discussed in this chapter have been extended and combined to produce ever more exotic varieties. In particular, early exercise can be built into almost any option. Examples can be found in (Hull, 2000; Kwok, 1998; Taleb, 1997; Wilmott, 1998).

Practical issues in the use of the Monte Carlo and binomial methods for exotic options are treated in (Clewlow and Strickland, 1998).

From a trader's perspective, 'how to hedge' is more important than 'how to value'. The hedging issue is covered in (Taleb, 1997; Wilmott, 1998).

EXERCISES

19.1. ★★★ Suppose that the function $V(S, t)$ satisfies the Black–Scholes PDE (8.15). Let

$$\widehat{V}(S, t) := S^{1 - \frac{2r}{\sigma^2}} V\left(\frac{X}{S}, t\right).$$

Show that

$$\frac{\partial \widehat{V}}{\partial t} + \tfrac{1}{2}\sigma^2 S^2 \frac{\partial^2 \widehat{V}}{\partial S^2} + rS\frac{\partial \widehat{V}}{\partial S} - r\widehat{V} = S^{1 - \frac{2r}{\sigma^2}}\left(\frac{\partial V}{\partial t}\left(\frac{X}{S}, t\right)\right.$$

$$+ \tfrac{1}{2}\sigma^2 \left(\frac{X}{S}\right)^2 \frac{\partial^2 V}{\partial S^2}\left(\frac{X}{S}, t\right) + r\frac{X}{S}\frac{\partial V}{\partial S}\left(\frac{X}{S}, t\right) - rV\left.\left(\frac{X}{S}, t\right)\right).$$

Deduce that $\widehat{V}(S, t)$ solves the Black–Scholes PDE.

19.2. ★★ Using Exercise 19.1, deduce that C^{B} in (19.3) satisfies the Black–Scholes PDE (8.15). Confirm also that C^{B} satisfies the conditions (19.1) and (19.2) when $B < E$.

19.3. ★★ Explain why (19.4) holds for all 'down' and 'up' barrier options.

19.4. ★ Why does it not make sense to have $B < E$ in an up-and-in call option?

19.5. ★★ The value of an up-and-out call option should approach zero as S approaches the barrier B from below. Verify that setting $S = B$ in (19.5) returns the value zero.

19.6. ★★★ Consider the geometric average price Asian call option, with payoff

$$\max\left[\left(\prod_{i=1}^{n} S(t_i)\right)^{1/n} - E, 0\right],$$

where the points $\{t_i\}_{i=1}^n$ are equally spaced with $t_i = i\,\Delta t$ and $n\,\Delta t = T$. By writing

$$\prod_{i=1}^n S(t_i) = \frac{S(t_n)}{S(t_{n-1})} \left(\frac{S(t_{n-1})}{S(t_{n-2})}\right)^2 \left(\frac{S(t_{n-2})}{S(t_{n-3})}\right)^3$$
$$\cdots \left(\frac{S(t_3)}{S(t_2)}\right)^{n-2} \left(\frac{S(t_2)}{S(t_1)}\right)^{n-1} \left(\frac{S(t_1)}{S_0}\right)^n S_0^n$$

and using the 'additive mean and variance' property of independent normal random variables mentioned as item (iii) at the end of Section 3.5, show that for the asset model (6.9) under risk neutrality, we have

$$\log\left[\left(\prod_{i=1}^n S(t_i)\right)^{1/n} \Big/ S_0\right] = \mathsf{N}\!\left((r - \tfrac{1}{2}\sigma^2)\frac{(n+1)}{2n}T, \sigma^2\frac{(n+1)(2n+1)}{6n^2}T\right).$$

(Note in particular that this establishes a lognormality structure, akin to that of the underlying asset.) Valuing the option as the risk-neutral discounted expected payoff, deduce that the time-zero option value is equivalent to the discounted expected payoff for a European call option whose asset has volatility $\widehat{\sigma}$ satisfying

$$\widehat{\sigma}^2 = \sigma^2\frac{(n+1)(2n+1)}{6n^2}$$

and drift $\widehat{\mu}$ given by

$$\widehat{\mu} = \tfrac{1}{2}\widehat{\sigma}^2 + (r - \tfrac{1}{2}\sigma^2)\frac{(n+1)}{2n}.$$

Use Exercise 12.4 and the Black–Scholes formula (8.19) to deduce that the time-zero geometric average price Asian call option value can be written

$$e^{-rT}\left(S_0 e^{\widehat{\mu}T} N(\widehat{d_1}) - EN(\widehat{d_2})\right), \tag{19.10}$$

where

$$\widehat{d_1} = \frac{\log(S_0/E) + (\widehat{\mu} + \tfrac{1}{2}\widehat{\sigma}^2)T}{\widehat{\sigma}\sqrt{T}},$$
$$\widehat{d_2} = \widehat{d_1} - \widehat{\sigma}\sqrt{T}.$$

19.7. ⋆⋆ Write down a pseudo-code algorithm for Monte Carlo applied to a floating strike lookback put option.

19.8 Program of Chapter 19 and walkthrough

In `ch19`, listed in Figure 19.4, we value an up-and-out call option. The first part of the code is a straightforward evaluation of the Black–Scholes formula (19.5). The second part shows how a Monte Carlo approach can be used. This code follows closely the algorithm outlined in Section 19.6, except that the asset path computation is *vectorized*: rather than loop for `j = 0 : N-1`, we compute the full path in one fell swoop, using the `cumprod` function that we encountered in `ch07`.

   Running `ch19` gives `bsval = 0.1857` for the Black–Scholes value and `conf = [0.1763, 0.1937]` for the Monte Carlo confidence interval.

PROGRAMMING EXERCISES

P19.1. Use a Monte Carlo approach to value a floating strike lookback put option.

P19.2. Implement the binomial method for a shout option, using (19.9), and investigate its rate of convergence.

Quotes

There are so many of them, and some of them are so esoteric,
that the risks involved may not be properly understood
even by the most sophisticated of investors.
Some of these instruments appear to be specifically designed to enable institutions
to take gambles which they would otherwise not be permitted to take ...
One of the driving forces behind the development of derivatives
was to escape regulations.

GEORGE SOROS, source (Bass, 1999)

The standard theory of contingent claim pricing through dynamic replication
gives no special role to options.
Using Monte Carlo simulation, path-dependent multivariate claims of great complexity
can be priced as easily as the path-independent univariate hockey-stick payoffs which
characterize options.
It is thus not at all obvious why markets have organized to offer these simple payoffs,
when other collections of functions
such as polynomials, circular functions, or wavelets
might offer greater advantages.

PETER CARR, KEITH LEWIS AND DILIP MADAN, 'On The Nature of Options',
Robert H. Smith School of Business, *Smith Papers Online*, 2001, source
http://bmgt1-notes.umd.edu/faculty/km/papers.nsf

Do you believe that huge losses on derivatives are confined to reckless or dim-witted
institutions?

```
%CH19        Program for Chapter 19
%
% Up-and-out call option
% Evaluates Black-Scholes formula and also uses Monte Carlo

randn('state',100)

%%%%%%%%% Problem and method parameters %%%%%%%%%%%
S = 5; E = 6; sigma = 0.25; r = 0.05; T = 1; B = 9;
Dt = 1e-3; N = T/Dt; M = 1e4;
%%%%%%%%%%%%%%%%%%%%%%%%%%%%%%%%%%%%%%%%%%%%%%

%%%%%%%%%%%%%%% Black-Scholes value %%%%%%%%%%%%%%%%%
tau = T;
power1 = -1 + (2*r)/(sigma^2);
power2 = 1 + (2*r)/(sigma^2);
d1 = (log(S/E) + (r + 0.5*sigma^2)*(tau))/(sigma*sqrt(tau));
d2 = d1 - sigma*sqrt(tau);
e1 = (log(S/B) + (r + 0.5*sigma^2)*(tau))/(sigma*sqrt(tau));
e2 = (log(S/B) + (r - 0.5*sigma^2)*(tau))/(sigma*sqrt(tau));
f1 = (log(S/B) - (r - 0.5*sigma^2)*(tau))/(sigma*sqrt(tau));
f2 = (log(S/B) - (r + 0.5*sigma^2)*(tau))/(sigma*sqrt(tau));
g1 = (log(S*E/(B^2)) - (r - 0.5*sigma^2)*(tau))/(sigma*sqrt(tau));
g2 = (log(S*E/(B^2)) - (r + 0.5*sigma^2)*(tau))/(sigma*sqrt(tau));
Nd1 = 0.5*(1+erf(d1/sqrt(2))); Nd2 = 0.5*(1+erf(d2/sqrt(2)));
Ne1 = 0.5*(1+erf(e1/sqrt(2))); Ne2 = 0.5*(1+erf(e2/sqrt(2)));
Nf1 = 0.5*(1+erf(f1/sqrt(2))); Nf2 = 0.5*(1+erf(f2/sqrt(2)));
Ng1 = 0.5*(1+erf(g1/sqrt(2))); Ng2 = 0.5*(1+erf(g2/sqrt(2)));
a = (B/S)^power1; b = (B/S)^power2;
bsval = S*(Nd1-Ne1-b*(Nf2-Ng2)) - E*exp(-r*tau)*(Nd2-Ne2-a*(Nf1-Ng1))
%%%%%%%%%%%%%%%%%%%%%%%%%%%%%%%%%%%%%%%%%%%%%%

V = zeros(M,1);
for i = 1:M
   Svals = S*cumprod(exp((r-0.5*sigma^2)*Dt+sigma*sqrt(Dt)*randn(N,1)));
   Smax = max(Svals);
   if Smax < B
     V(i) = exp(-r*T)*max(Svals(end)-E,0);
   end
end
aM = mean(V); bM = std(V);
conf = [aM - 1.96*bM/sqrt(M), aM + 1.96*bM/sqrt(M)]
```

Fig. 19.4.   Program of Chapter 19: ch19.m.

If so, consider:

Procter & Gamble (lost $102 million in 1994)

Gibson Greetings (lost $23 million in 1994)

Orange County, California (bankrupted after $1.7 billion loss in 1994)

Baring's Bank (bankrupted after $1.3 billion loss in 1995)

Sumitomo (lost $1.3 billion in 1996)

Government of Belgium ($1.2 billion loss in 1997)

National Westminster Bank (lost $143 million in 1997)

PHILIP MCBRIDE JOHNSON (Johnson, 1999)

20

Historical volatility

20.1 Motivation

We know that the volatility parameter, σ, in the Black–Scholes formula cannot be observed directly. In Chapter 14 we saw how σ for a particular asset can be estimated as the *implied volatility*, based on a reported option value. In this chapter we discuss another widely used approach – estimating the volatility from the previous behaviour of the asset. This technique is independent of the option valuation problem. Here is the basic principle.

Given that we have (a) a model for the behaviour of the asset price that involves σ and (b) access to asset prices for all times up to the present, let us fit σ in the model to the observed data.

A value $\sigma^\star$ arising from this general procedure is called a *historical volatility* estimate.

20.2 Monte Carlo type estimates

We suppose that historical asset price data is available at equally spaced time values $t_i := i \Delta t$, so $S(t_i)$ is the asset price at time t_i. We then define the log ratios

$$U_i := \log \frac{S(t_i)}{S(t_{i-1})}. \tag{20.1}$$

Our asset price model (6.9) assumes that the $\{U_i\}$ are independent, normal random variables with mean $(\mu - \frac{1}{2}\sigma^2)\Delta t$ and variance $\sigma^2 \Delta t$. From this point of view, getting hold of historical asset price data and forming the log ratios is equivalent to sampling from an $\mathsf{N}((\mu - \frac{1}{2}\sigma^2)\Delta t, \sigma^2 \Delta t)$ distribution. Hence, we could use a Monte Carlo approach to estimate the mean and variance. Suppose that $t = t_n$ is the current time and that the $M + 1$ most current asset prices $\{S(t_{n-M}), S(t_{n-M+1}), \ldots, S(t_{n-1}), S(t_n)\}$ are available. Using the corresponding log ratio data, $\{U_{n+1-i}\}_{i=1}^{M}$, the sample mean (15.1) and variance estimate (15.2) become

$$a_M := \frac{1}{M} \sum_{i=1}^{M} U_{n+1-i}, \tag{20.2}$$

$$b_M^2 := \frac{1}{M-1} \sum_{i=1}^{M} (U_{n+1-i} - a_M)^2. \tag{20.3}$$

We may therefore estimate the unknown parameter σ by comparing the sample mean a_M with the exact mean $(\mu - \frac{1}{2}\sigma^2)\Delta t$ from the model, or by comparing the sample variance b_M^2 with the exact variance $\sigma^2 \Delta t$ from the model. In practice the latter works much better – see Exercise 20.1 – and hence we let

$$\sigma^{\star} := \frac{b_M}{\sqrt{\Delta t}}. \tag{20.4}$$

Exercise 20.2 shows that this can be written directly in terms of the U_i values as

$$\sigma^{\star} = \sqrt{\frac{1}{\Delta t} \left(\frac{1}{M-1} \sum_{i=1}^{M} U_{n+1-i}^2 - \frac{1}{M(M-1)} \left(\sum_{i=1}^{M} U_{n+1-i} \right)^2 \right)}. \tag{20.5}$$

20.3 Accuracy of the sample variance estimate

To get some idea of the accuracy of the estimate $\sigma^{\star}$ in (20.4) we take the view that we are essentially using Monte Carlo simulation to compute b_M^2 as an approximation to the expected value of the random variable $(U - \mathbb{E}(U))^2$, where $U \sim \mathsf{N}((\mu - \frac{1}{2}\sigma^2)\Delta t, \sigma^2 \Delta t)$. (This is not exactly the case, as we are using an approximation to $\mathbb{E}(U)$.) Equivalently, after dividing through by Δt, we are using Monte Carlo simulation to compute $\sigma^{\star 2} = b_M^2/\Delta t$ as an approximation to the expected value of the random variable $(\widehat{U} - \mathbb{E}(\widehat{U}))^2$, where $\widehat{U} \sim \mathsf{N}((\mu - \frac{1}{2}\sigma^2)\sqrt{\Delta t}, \sigma^2)$. Hence, from (15.5), an approximate 95% confidence interval for σ^2 is given by

$$\sigma^{\star 2} \pm \frac{1.96 v}{\sqrt{M}},$$

where v^2 is the variance of the random variable $(\widehat{U} - \mathbb{E}(\widehat{U}))^2$. Exercise 20.3 shows that

$$v^2 = 2\sigma^4. \tag{20.6}$$

So the approximate confidence interval for σ^2 has the form

$$\sigma^{\star 2} \pm \frac{1.96\sqrt{2}\sigma^2}{\sqrt{M}}. \tag{20.7}$$

It may then be argued that

$$\sigma^\star \pm \frac{1.96\sigma^\star}{\sqrt{2M}} \tag{20.8}$$

is an approximate 95% confidence interval for σ, see Exercise 20.4. In particular, we recover the usual $1/\sqrt{M}$ behaviour.

There is, however, a subtle point to be made. In a typical Monte Carlo simulation, taking more samples (increasing M) means making more calls to a pseudo-random number generator. In the above context, though, taking more samples means looking up more data. There are two natural ways to do this.

(1) Keep Δt fixed and simply go back further in time.
(2) Fix the time interval, $M\Delta t$, over which the data is sampled and decrease Δt.

Both approaches are far from perfect. Case (1) runs counter to the intuitive notion that recent data is more important than old data. (The asset price yesterday is more relevant than the asset price last year.) We will return to this issue later. Case (2) suffers from a practical limitation: the bid–ask spread introduces a noisy component into the asset price data that becomes significant when very small Δt values are measured. Overall, finding a compromise between large M and small Δt is a difficult task.

If $\sigma^\star$ is computed in order to value an option, then a widely quoted rule of thumb is to make the historical data time-frame $M\Delta t$ equal to that of the option: to value an option that expires in six months' time, take six months of historical data. There is also some evidence that taking longer historical data periods is worthwhile.

Using the identity $\log(a/b) = \log a - \log b$ to simplify (20.2) we find that

$$a_M = \frac{1}{M} \sum_{i=1}^{M} (\log S(t_{n+1-i}) - \log S(t_{n-i}))$$

$$= \frac{1}{M} (\log S(t_n) - \log S(t_{n-M}))$$

$$= \frac{1}{M} \log \frac{S(t_n)}{S(t_{n-M})}.$$

Because those intermediate terms cancel, a_M depends only on the first and last S values! Our asset price model assumes that $\log(S(t_n)/S(t_{n-M}))$ is normal with mean $(\mu - \frac{1}{2}\sigma^2)M\,\Delta t$ and variance $\sigma^2 M\,\Delta t$. Hence,

$$a_M \sim \mathsf{N}\left((\mu - \tfrac{1}{2}\sigma^2)\Delta t, \sigma^2\frac{\Delta t}{M}\right). \tag{20.9}$$

In practice, because a_M is normal with small mean and variance, it is common to replace it by zero in (20.3), which leads to

$$\sigma^\star = \sqrt{\frac{1}{\Delta t}\frac{1}{(M-1)}\sum_{i=1}^{M} U_{n+1-i}^2}, \tag{20.10}$$

instead of (20.5). This alternative has been found to be more reliable in general.

20.4 Maximum likelihood estimate

To justify further the historical volatility estimate (20.10), we will show that an almost identical quantity

$$\sigma^\star = \sqrt{\frac{1}{\Delta t}\frac{1}{M}\sum_{i=1}^{M} U_{n+1-i}^2} \tag{20.11}$$

can be derived from a *maximum likelihood* viewpoint. Note that (20.11) differs from (20.10) only in that $M - 1$ has become M.

The maximum likelihood principle is based on the following idea:

In the absence of any extra information, assume the event that we observed was the one that was most likely to happen.

In terms of fitting an unknown parameter, the idea becomes:

Choose the parameter value that makes the event that we observed have the maximum probability.

As a simple example, consider the case where a coin is flipped four times. Suppose we think the coin is potentially biased – there is some $p \in [0, 1]$ such that, independently on each flip, the probability of heads (H) is p and the probability of tails (T) is $1 - p$. Suppose the four flips produce H,T,T,H. Then, under our assumption, the probability of this outcome is $p \times (1 - p) \times (1 - p) \times p = p^2(1 - p)^2$. Simple calculus shows that maximizing $p^2(1 - p)^2$ over $p \in [0, 1]$ leads to $p = \frac{1}{2}$, which is, of course, intuitively reasonable for that data. Similarly, if we observed H,T,H,H, the resulting probability is $p^3(1 - p)$. In this case, maximizing over $p \in [0, 1]$ gives $p = \frac{3}{4}$, also agreeing with our intuition.

That simple example involved a sequence of independent observations, where each observation (the result of a coin flip) is a discrete random variable. In the case where the model involves outcomes, say $U_1, U_2, \ldots, U_M$, from a continuous random variable with density $f(x)$ that involves some parameter, we look for the parameter value that maximizes the product

$$f(U_1) f(U_2) \ldots f(U_M).$$

Formally, this maximizes the value of the corresponding probability density function at the point $(U_1, U_2, \ldots, U_M)$.

Returning to the case of estimating the value σ from our observations of U_i in (20.1), we first make a simplification. On the basis that $U_i/\sqrt{\Delta t} \sim N((\mu - \frac{1}{2}\sigma^2)\sqrt{\Delta t}, \sigma^2)$, we take the view that the mean of $U_i/\sqrt{\Delta t}$ is negligible, and regard it as zero. The corresponding density function for each scaled observation $U_i/\sqrt{\Delta t}$ becomes

$$f(x) = \frac{1}{\sqrt{2\pi\sigma^2}} e^{-x^2/(2\sigma^2)}.$$

Our maximum likelihood estimate of σ is then found by maximizing

$$\prod_{i=1}^{M} \frac{1}{\sqrt{2\pi\sigma^2}} e^{-\widehat{U}_{n+1-i}^2/(2\sigma^2)}, \qquad \text{where } \widehat{U}_{n+1-i} = U_{n+1-i}/\sqrt{\Delta t}. \tag{20.12}$$

Exercise 20.5 shows that this leads to the estimate (20.11).

20.5 Other volatility estimates

Under the simplifying assumption that U_i has zero mean, $\mathrm{var}(U_i) = \mathbb{E}(U_i^2)$. For the estimate (20.11) we have

$$\Delta t \sigma^{\star 2} = \frac{1}{M} \sum_{i=1}^{M} U_{n+1-i}^2,$$

and hence we may interpret $\Delta t \sigma^{\star 2}$ as a sample mean approximation for this expected value. Keep in mind that the samples, U_i^2, correspond to different points in time. It has been found that rather than treating each observation U_i equally it is more appropriate to give extra weight to the most recent values. This leads to schemes of the general form

$$\Delta t \sigma^{\star 2} = \sum_{i=1}^{M} \alpha_i U_{n+1-i}^2, \qquad \text{where } \sum_{i=1}^{M} \alpha_i = 1, \tag{20.13}$$

with $\alpha_1 > \alpha_2 > \cdots > \alpha_M > 0$. It is common to use geometrically declining weights: $\alpha_{i+1} = w\alpha_i$, for some $0 < w < 1$. This produces the estimate

$$\Delta t \sigma^{\star 2} = \frac{\sum_{i=1}^{M} w^i U_{n+1-i}^2}{\sum_{i=1}^{M} w^i}.$$

The choice $w = 0.94$ is popular. Note that $(0.94)^{10} \approx 0.54$, $(0.94)^{100} \approx 0.0021$ and $(0.94)^{200} < 10^{-5}$, so in this case, even if M is chosen to be very large, samples more than around a hundred Δt units old are essentially ignored.

If a new volatility estimate is needed at each time t_n, there is a neat variation of this idea. Suppose $\Delta t \sigma_n^{\star 2}$ is our estimate of $\Delta t \sigma^2$ computed at time t_n, based on $\{U_{n+1-i}\}_{i=1}^{M}$. Then an estimate for time t_{n+1} can be computed as

$$\Delta t \sigma_{n+1}^{\star 2} = w \Delta t \sigma_n^{\star 2} + (1 - w) U_{n+1}^2. \tag{20.14}$$

This process is close to having geometrically declining weights, see Exercise 20.7, and has the advantage that updating from time t_n to time t_{n+1} does not require the old data U_i, for $i \leq n$, to be accessed.

Formulas such as (20.14) are sometimes referred to as exponentially weighted moving average (EWMA) models. Of course, the notion of computing a time-varying estimate of the volatility is inherently at odds with the underlying assumption of constant volatility that is used in the derivation of the Black–Scholes formula. Even so, it has been observed empirically that asset price volatility is not constant, and techniques that account for this fact have proved successful.

20.6 Example with real data

In Figure 20.1 we estimate historical volatility for the IBM daily and weekly data from Figures 5.1 and 5.2. In both cases, we assume that the data corresponds to equally spaced points in time. The daily data runs over 9 months ($T = 3/4$ years) and has 183 asset prices ($M = 182$), so we set $\Delta t = T/M \approx 0.0041$. The weekly data runs over 4 years ($T = 4$) and has 209 asset prices ($M = 208$), so we set $\Delta t = T/M \approx 0.0192$.

For the daily data we found $a_M = -4.3 \times 10^{-4}$, confirming that it is reasonable to regard the log ratio mean as zero. The Monte Carlo based estimate (20.4) produced $\sigma^\star = 0.4069$ with a 95% confidence interval of $[0.3653, 0.4486]$. Given that $a_M \approx 0$, it is not surprising that the simpler estimate (20.10) produced an almost identical value $\sigma^\star = 0.4070$. This $\sigma^\star$ is represented as a dashed line in the upper picture. The EWMA is plotted as diamond shaped markers joined by straight lines. Here, we used the first 20 U_i values to compute a Monte Carlo based estimate, and inserted this as a starting value for $\sigma^\star$ in the update formula (20.14). Our weight was $w = 0.94$.

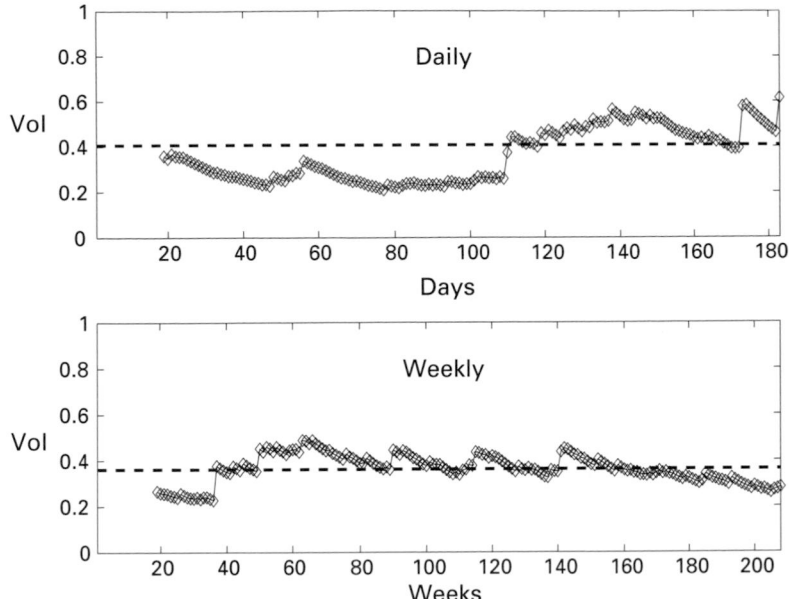

Fig. 20.1. Historical volatility estimates for IBM data from Figures 5.1 and 5.2. Upper picture: daily. Lower picture: weekly. Diamonds are exponentially weighted moving averages. Dashed lines show the estimate (20.10).

The lower picture repeats the exercise for the weekly data. In this case (20.4) produced $\sigma^\star = 0.3610$ with a 95% confidence interval of $[0.3263, 0.3957]$. We found that $a_M = -4.0 \times 10^{-3}$ and the estimate (20.10) gave $\sigma^\star = 0.3621$.

Overall, the small size of the sample mean a_M and the reasonable agreement between the daily and weekly $\sigma^\star$ estimates are encouraging. However, the large confidence intervals for these estimates, and the significant time dependency of the EWMA, are far from reassuring. Generally, extracting historical volatility estimates from real data is a mixture of art and science.

20.7 Notes and references

Volatility estimation is undoubtedly one of the most important aspects of practical option valuation, and it remains an active research topic, see (Poon and Granger, 2003), for example.

More sophisticated time-varying volatility models, including autoregressive conditional heteroscedasticity (ARCH) and generalized autoregressive conditional heteroscedasticity (GARCH) are discussed in (Hull, 2000), for example.

In addition to providing information for option valuation, historical volatility estimates are a key component in the determination of *Value at Risk*; see (Hull, 2000, Chapter 4), for example.

20.1. ⋆⋆⋆ Consider a Monte Carlo approach where the sample mean a_M in (20.2) is used to approximate the exact mean $(\mu - \frac{1}{2}\sigma^2)\Delta t$ in order to estimate σ. Suppose that a fixed time-frame $M\Delta t$ is used for the log ratios. This corresponds to case (2) in Section 20.3. Show that the 95% confidence interval for the mean has width proportional to $1/M$. Convince yourself that this is a poor method. [Hint: use (20.9) and refer to Chapter 15.]

20.2. ⋆⋆ Establish (20.5).

20.3. ⋆⋆ Let $Z \sim N(0, 1)$ and $Y = \alpha + \beta Z$, for $\alpha, \beta \in \mathbb{R}$. Show that $\text{var}\left((Y - \mathbb{E}(Y))^2\right) = 2\beta^4$. Hence, verify (20.6).

20.4. ⋆⋆ Use the expansion $\sqrt{1 \pm \epsilon} \approx 1 \pm \frac{1}{2}\epsilon$ for small $\epsilon > 0$, to show how the approximate confidence interval (20.8) may be inferred from (20.7).

20.5. ⋆⋆⋆ Show that maximizing (20.12) with respect to σ leads to the estimate (20.11). [Hints: (1) take logs – maximizing a positive quantity is equivalent to maximizing its log, (2) regard σ^2 as the unknown parameter, rather than σ.]

20.6. ⋆ Give a convincing argument for the constraint $\sum_{i=1}^{M} \alpha_i = 1$ in (20.13).

20.7. ⋆⋆ Explain why (20.14) almost corresponds to having geometrically declining weights.

20.8 Program of Chapter 20 and walkthrough

In ch20, listed in Figure 20.2, we look at historical volatility estimation with artificial data, created with a random number generator. The array U has ith entry given by the log of the ratio asset(i+1)/asset(i), where the asset path, asset, is created in the usual way, using a volatility of sigma = 0.3. The Monte Carlo volatility estimate (20.4) turns out to be 0.2947, with an approximate confidence interval (20.8) of cont = [0.2855, 0.3038]. The simplified estimate with sample mean set to zero also gives sigma2 = 0.2947. We then apply the EWMA formula (20.14) with $w = 0.94$, using a Monte Carlo estimate of the first twenty U values to initialize the volatility. The running estimate, s, is plotted and the exact level 0.3 is superimposed as a dashed white line. Figure 20.3 shows the picture. The final time EWMA volatility value was sigma3 = 0.2588.

In this example, the Monte Carlo version performs better than EWMA. This is to be expected – we are generating paths that agree with our underlying model (6.9), so taking as many old data points as possible is clearly a good idea. The EWMA approach of giving extra weight to more recent data points is designed to improve the estimate when real stock market data is used.

P20.1. Apply the techniques in ch20 to some real option data.

P20.2. Compare implied and historical volatility estimates on some real option data.

```
%CH20      Program for Chapter 20
%
% Computes historical volatility from artificially generated data

clf
randn('state',100)

%%%%%%%%%%% Parameters %%%%%%%%%%%%%%%
sigma = 0.3; r = 0.03; M = 2e3; Dt = 1/(M+1);
%%%%%%%%%%%%%%%%%%%%%%%%%%%%%%%%%%%%%%%

asset = cumprod(exp((r-0.5*sigma^2)*Dt+sigma*sqrt(Dt)*randn(M+1,1)));
U = log(asset(2:end)./asset(1:end-1));

%% Monte Carlo estimate based on all data %%
Umean = mean(U);
Ustd = std(U);
sigma1 = Ustd/sqrt(Dt)
cont = [sigma1*(1-1.96/sqrt(2*M)), sigma1*(1+1.96/sqrt(2*M))]

%% Simplified estimate (assumes zero mean)
sigma2 = sqrt(sum(U.^2)/((M-1)*Dt))

%% Running EWMA %%
%% First get a starting value %%
s = zeros(M,1);
L = 20;
V = U(1:L);
s(L) = std(V)/sqrt(Dt);
%% Now do EWMA %%
w = 0.94;
for n = L:M-1
    s(n+1) = sqrt((w*Dt*s(n)^2 + (1-w)*U(n+1)^2)/Dt);
end
sigma3 = s(end)

plot([L:M], s(L:end),'r-d')
hold on
plot([1 M],[sigma sigma],'w--','LineWidth',2)
xlabel('t'), ylabel('Volatility'), ylim([0, 0.5]), grid on
```

Fig. 20.2.   Program of Chapter 20: ch20.m.

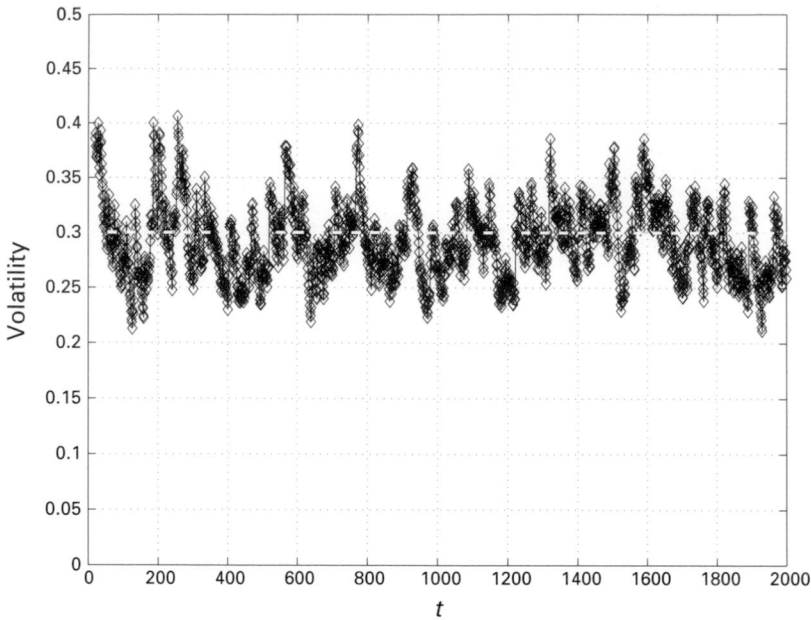

Fig. 20.3.   Figure produced by ch20.

Quotes

There are two main approaches to estimating volatility and correlation:
a direct approach using historical data
and an indirect approach of inferring volatility from option prices.
The historical approach has the virtue of working directly
with the most relevant data but is always handicapped by 'looking backward'.
Implied volatility is a naturally forward-looking measure,
but it is difficult to separate estimation error from model error.
For example, differing Black–Scholes implied volatilities
could be due to non-constant volatility
or could be due to violations of perfect market assumptions
that have unequal impacts on different options
(e.g., differences in liquidity and transactions costs among options).
> MARK BROADIE AND PAUL GLASSERMAN (Broadie and Glasserman, 1998)

A headline in Enron's 2000 annual report states
'In Volatile Markets, Everything Changes But Us.'
Sadly, Enron got it wrong.
> Testimony of FRANK PARTNOY, Professor of Law, University of San Diego School of
> Law, hearings before the United States Senate Committee on Governmental Affairs, 24
> January 2002. Taken from *Financial Engineering News*, June/July 2002, Issue No. 26.

Since the statistical properties of the sample mean
make it a very inaccurate estimate of the true mean,

taking deviations around zero rather than around the sample mean
typically increases forecast accuracy.

T. CLIFTON GREEN AND STEPHEN FIGLEWSKI (Green and Figlewski, 1999)

The authors emphasize that,
as even the most cursory examination of the historical record reveals,
'geometric Brownian motion' is at best a first approximation to the actual
movements of the price of any real stock or collection of stocks.
Even their assumption that the governing processes
are stochastic – rather than examples of deterministic chaos – may in time
be disproved by sufficiently sensitive measurement techniques.

JAMES CASE, reviewing the book (Mantegna and Stanley, 2000) in *Society for Industrial and Applied Mathematics (SIAM) News*, Volume 34, January/February 2001.

Long run is a misleading guide to current affairs.
In the long run we are all dead.

JOHN MAYNARD KEYNES (1883–1946), *A Treatise on Monetary Reform*, Chapter 3, 1923.

21

Monte Carlo Part II: variance reduction by antithetic variates

21.1 Motivation

The Monte Carlo method gives a simple and flexible technique for option valuation. However, we have seen that it can be expensive. This chapter and the next cover two approaches that attempt to improve efficiency. The antithetic variates idea in this chapter has the benefit of being widely applicable and easy to implement. In order to understand how the idea works, we need to discuss the concept of covariance between random variables.

21.2 The big picture

The Monte Carlo method uses the sample mean (15.1) to approximate the expected value of the random variable X, where the X_i are i.i.d. with $\mathbb{E}(X_i) = \mathbb{E}(X)$. We saw in Chapter 15 that the width of the corresponding confidence interval is inversely proportional to $\sqrt{M}$. This makes it an expensive business to improve the approximation by taking more samples. To get an extra digit of accuracy, that is, to shrink a confidence interval by a factor of 10, requires 100 times as many samples. However, we also saw that the confidence interval width scales with $\sqrt{\text{var}(X_i)}$. This motivates the idea of replacing the X_i in (15.1) with another sequence of i.i.d. random variables that have the same mean as the X_i but with smaller variance. This is the idea behind *variance reduction*. One way to summarize the potential advantage is:

If we can reduce the variance in X_i by a factor $R < 1$, then for a given number of samples, M, the new version has confidence intervals that are a factor $\sqrt{R}$ smaller. So for $R = 10^{-k}$ the new version gives roughly $k/2$ extra digits of accuracy.

Under the assumption that sampling from the new random variable sequence costs about the same as sampling from X_i, we could re-state this from a slightly different viewpoint:

If we can reduce the variance in X_i by a factor $R < 1$, then the new method gives confidence intervals of the same width for R times less work.

21.3 Dependence

So far, we have focused on collections of *independent* random variables. In particular, we have repeatedly used the result (3.9): if X and Y are independent then $\mathbb{E}(XY) = \mathbb{E}(X)\mathbb{E}(Y)$. To discuss variance reduction techniques for Monte Carlo, we now need to consider the case where random variables are not independent.

Intuitively, two random variables are independent if knowing the value of one does not give any information about the value of the other. For illustration, suppose we flip two coins. Denote the four possible outcomes by $\{H_1, H_2\}$, $\{H_1, T_2\}$, $\{T_1, H_2\}$, $\{T_1, T_2\}$, where, for example, $\{H_1, T_2\}$ signifies heads for the first coin and tails for the second. Now define random variables X and Y as follows. Let X take the value 1 if the first coin lands heads and 0 otherwise, and let Y take the value 1 if the first *and* second coins land heads and 0 otherwise. Thus

$$X = \begin{cases} 1, & \text{for } \{H_1, H_2\}, \\ 1, & \text{for } \{H_1, T_2\}, \\ 0, & \text{for } \{T_1, H_2\}, \\ 0, & \text{for } \{T_1, T_2\}, \end{cases} \quad \text{and} \quad Y = \begin{cases} 1, & \text{for } \{H_1, H_2\}, \\ 0, & \text{for } \{H_1, T_2\}, \\ 0, & \text{for } \{T_1, H_2\}, \\ 0, & \text{for } \{T_1, T_2\}. \end{cases}$$

It is intuitively clear that X and Y are not independent – for example, knowing that $X = 0$ allows us to deduce immediately that $Y = 0$. To work out the expected values $\mathbb{E}(X)$, $\mathbb{E}(Y)$ and $\mathbb{E}(XY)$, we need the following information.

	Probability	X	Y	XY
$\{H_1, H_2\}$	$\frac{1}{4}$	1	1	1
$\{H_1, T_2\}$	$\frac{1}{4}$	1	0	0
$\{T_1, H_2\}$	$\frac{1}{4}$	0	0	0
$\{T_1, T_2\}$	$\frac{1}{4}$	0	0	0

So applying the expected value formula (3.1), $\mathbb{E}(X) = \frac{1}{2}$, $\mathbb{E}(Y) = \frac{1}{4}$ and $\mathbb{E}(XY) = \frac{1}{4}$. Thus we have $\mathbb{E}(XY) \neq \mathbb{E}(X)\mathbb{E}(Y)$, confirming that X and Y cannot be independent.

As a measure of 'dependence' between the random variables X and Y, the *covariance*, $\mathsf{cov}(X, Y)$, is defined as follows,

$$\mathsf{cov}(X, Y) := \mathbb{E}\left[(X - \mathbb{E}(X))(Y - \mathbb{E}(Y))\right]. \tag{21.1}$$

Equivalently, we may write

$$\mathsf{cov}(X, Y) := \mathbb{E}(XY) - \mathbb{E}(X)\mathbb{E}(Y), \tag{21.2}$$

see Exercise 21.1, and it follows immediately that if X and Y are independent then $\mathsf{cov}(X, Y) = 0$. Loosely, from (21.1), if the covariance is positive then X and Y tend to be smaller than their means or larger than their means at the same time. Similarly, if the covariance is negative then X tends to be above its mean when Y is below its mean, and vice versa. In the example above we have $\mathsf{cov}(X, Y) = \frac{1}{8}$, which supports this interpretation.

21.4 Antithetic variates: uniform example

To illustrate the use of antithetic variates, suppose we apply Monte Carlo to approximate the expected value

$$I = \mathbb{E}(e^{\sqrt{U}}), \qquad \text{where } U \sim \mathsf{U}(0, 1). \tag{21.3}$$

For reference, we note that $I = 2$, see Exercise 21.2. A Monte Carlo estimate of I is

$$I_M = \frac{1}{M} \sum_{i=1}^{M} Y_i, \qquad \text{where } Y_i = e^{\sqrt{U_i}} \text{ with i.i.d. } U_i \sim \mathsf{U}(0, 1). \tag{21.4}$$

We know that the accuracy of Monte Carlo is related to the variance of Y_i. In this case we have

$$\mathsf{var}(Y_i) = \frac{e^2 - 7}{2}, \tag{21.5}$$

see Exercise 21.2.

Now consider using the *antithetic variate* Monte Carlo estimator

$$\widehat{I}_M = \frac{1}{M} \sum_{i=1}^{M} \widehat{Y}_i, \qquad \text{where } \widehat{Y}_i = \frac{e^{\sqrt{U_i}} + e^{\sqrt{1-U_i}}}{2}, \qquad \text{with } U_i \sim \mathsf{U}(0, 1). \tag{21.6}$$

This antithetic version 're-uses' U_i in the form $1 - U_i$. Note that $1 - U_i \sim \mathsf{U}(0, 1)$, so $\mathbb{E}(\widehat{Y}_i) = \mathbb{E}(e^{\sqrt{U}})$. In terms of random number generation, I_M and $\widehat{I}_M$

have the same costs – both use M uniform samples. (Note, however, that there is some overhead associated with $\widehat{I}_M$. Twice as many evaluations of the exponential and square root functions are required.)

From the useful identity

$$\text{var}(X + Y) = \text{var}(X) + \text{var}(Y) + 2\,\text{cov}(X, Y), \qquad (21.7)$$

see Exercise 21.3, we have

$$\text{var}\left(\frac{e^{\sqrt{U_i}} + e^{\sqrt{1-U_i}}}{2}\right) = \tfrac{1}{4}\left(\text{var}(e^{\sqrt{U_i}}) + \text{var}(e^{\sqrt{1-U_i}}) + 2\,\text{cov}(e^{\sqrt{U_i}}, e^{\sqrt{1-U_i}})\right)$$

$$= \tfrac{1}{2}\left(\text{var}(e^{\sqrt{U_i}}) + \text{cov}(e^{\sqrt{U_i}}, e^{\sqrt{1-U_i}})\right). \qquad (21.8)$$

This is the key expression that tells us how the variance in $\widehat{Y}_i$ compares with the variance in the original Y_i.

Now

$$\mathbb{E}\left(e^{\sqrt{U_i}} e^{\sqrt{1-U_i}}\right) = \int_0^1 e^{\sqrt{x}+\sqrt{1-x}}\,dx$$

and hence, using (21.2),

$$\text{cov}\left(e^{\sqrt{U_i}}, e^{\sqrt{1-U_i}}\right) = \int_0^1 e^{\sqrt{x}+\sqrt{1-x}}\,dx - 2 \times 2. \qquad (21.9)$$

Inserting (21.5) and (21.9) in (21.8), we arrive at

$$\text{var}(\widehat{Y}_i) = \frac{e^2}{4} - \frac{15}{4} + \frac{1}{2}\int_0^1 e^{\sqrt{x}+\sqrt{1-x}}\,dx. \qquad (21.10)$$

Using a numerical quadrature routine to approximate the integral in (21.10), we find that

$$\text{var}(\widehat{Y}_i) \approx 0.001\,073.$$

Hence,

$$\frac{\text{var}(Y_i)}{\text{var}(\widehat{Y}_i)} \approx 181.2485. \qquad (21.11)$$

It follows from the discussion in Section 21.2 that the antithetic version gives us at least an extra digit of accuracy for the same amount of random number generation.

Computational example Table 21.1 shows the 95% confidence intervals for I_M and the antithetic version $\widehat{I}_M$ for the problem (21.3). We did four tests, covering $M = 10^2, 10^3, 10^4, 10^5$, and used the same random number samples for the two methods. In addition to the confidence intervals, we give the ratio of the sizes

Table 21.1. *Ninety-five per cent confidence intervals for (21.4) and (21.6) on problem (21.3), plus ratios of their widths*

M	Standard	Antithetic	Ratio of widths
10^2	[1.8841, 2.0752]	[1.9875, 2.0012]	14.0
10^3	[1.9538, 2.0087]	[1.9976, 2.0017]	13.4
10^4	[1.9890, 2.0062]	[1.9997, 2.0010]	13.5
10^5	[1.9969, 2.0023]	[1.9998, 2.0002]	13.5

of the two confidence intervals. This is precisely the ratio of the square roots of the sample variances. As predicted by (21.11), it converges to $\sqrt{181.2485} \approx 13.5$. As a practical note, it is worth emphasizing that the confidence intervals for the antithetic variates estimate were computed via the sample variance of $\{\widehat{Y}_i\}_{i=1}^M$, which are independent, and **not** $\{e^{\sqrt{U_i}}\}_{i=1}^M \cup \{e^{\sqrt{1-U_i}}\}_{i=1}^M$, which are highly correlated. ◇

21.5 Analysis of the uniform case

To understand how the antithetic variate technique works, consider the more general case of approximating

$$I = \mathbb{E}(f(U)), \qquad \text{where } U \sim \mathsf{U}(0, 1),$$

for some function f. The standard Monte Carlo estimate is

$$I_M = \frac{1}{M} \sum_{i=1}^M f(U_i), \qquad \text{with i.i.d. } U_i \sim \mathsf{U}(0, 1), \qquad (21.12)$$

and the antithetic alternative is

$$\widehat{I}_M = \frac{1}{M} \sum_{i=1}^M \frac{f(U_i) + f(1 - U_i)}{2}, \qquad \text{with i.i.d. } U_i \sim \mathsf{U}(0, 1). \qquad (21.13)$$

Copying the way that we derived (21.8), we find that

$$\mathsf{var}\left(\frac{f(U_i) + f(1 - U_i)}{2}\right) = \tfrac{1}{2}\left(\mathsf{var}\left(f(U_i)\right) + \mathsf{cov}\left(f(U_i), f(1 - U_i)\right)\right). \tag{21.14}$$

The success of the new scheme hinges on whether $\mathsf{var}(\frac{1}{2}(f(U_i) + f(1 - U_i)))$ is smaller than $\mathsf{var}(f(U_i))$. The identity (21.14) tells us that efficiency boils down to making $\mathsf{cov}(f(U_i), f(1 - U_i))$ *as negative as possible*. We want $f(U_i)$ to be big (relative to its mean) when $f(1 - U_i)$ is small (relative to its mean). Intuitively, this approach will work when f is monotonic. Loosely, the antithetic variate technique attempts to compensate for samples that are above the mean by adding samples that are below the mean, and vice versa.

We may convert this intuition into a mathematical result. First we recall that to say a function f is *monotonic increasing* means $x_1 \leq x_2 \Rightarrow f(x_1) \leq f(x_2)$. Similarly, to say a function f is *monotonic decreasing* means $x_1 \leq x_2 \Rightarrow f(x_1) \geq f(x_2)$. It follows straightforwardly that if f and g are both monotonic increasing functions or both monotonic decreasing functions then

$$(f(x) - f(y))(g(x) - g(y)) \geq 0, \qquad \text{for any } x \text{ and } y, \qquad (21.15)$$

see Exercise 21.5. Now we prove a useful lemma.

Lemma If f and g are both monotonic increasing functions or both monotonic decreasing functions then, for any random variable X,

$$\mathsf{cov}(f(X), g(X)) \geq 0.$$

Proof   Let Y be a random variable that is independent of X with the same distribution. From (21.15) we may write

$$(f(X) - f(Y))(g(X) - g(Y)) \geq 0.$$

So the random variable $(f(X) - f(Y))(g(X) - g(Y))$ must have a non-negative expected value. Hence

$$0 \leq \mathbb{E}[(f(X) - f(Y))(g(X) - g(Y))]$$
$$= \mathbb{E}[f(X)g(X)] - \mathbb{E}[f(X)g(Y)] - \mathbb{E}[f(Y)g(X)] + \mathbb{E}[f(Y)g(Y)].$$

Since X and Y are i.i.d., that last right-hand side simplifies to

$$2\mathbb{E}[f(X)g(X)] - 2\mathbb{E}[f(X)]\mathbb{E}[g(X)],$$

which is $2\,\mathsf{cov}(f(X), g(X))$, and the result follows.   ◇

Now note that if f is a monotonic increasing function, then so is $-f(1 - x)$. Similarly, if f is a monotonic decreasing function, then so is $-f(1 - x)$. In either case, applying our lemma gives $\mathsf{cov}(f(X), -f(1 - X)) \geq 0$. Equivalently, $\mathsf{cov}(f(X), f(1 - X)) \leq 0$. In (21.14) this shows that

$$\mathsf{var}\left(\frac{f(U_i) + f(1 - U_i)}{2}\right) \leq \tfrac{1}{2}\mathsf{var}(f(U_i)), \qquad (21.16)$$

when f is monotonic. In words:

For monotonic f, the variance in the antithetic sample is always less than or equal to half that in the standard sample.

Of course, this is only a bound. The actual improvement can be much better, as in the $f(x) = e^{\sqrt{x}}$ example of the previous section.

21.6 Normal case

The antithetic variates trick is not restricted to functions of uniform random variables. In the case of

$$I = \mathbb{E}(f(U)), \qquad \text{where } U \sim \mathsf{N}(0, 1), \qquad (21.17)$$

the standard Monte Carlo estimate is

$$I_M = \frac{1}{M} \sum_{i=1}^{M} f(U_i), \qquad \text{with i.i.d. } U_i \sim \mathsf{N}(0, 1), \qquad (21.18)$$

and the antithetic alternative is

$$\widehat{I}_M = \frac{1}{M} \sum_{i=1}^{M} \frac{f(U_i) + f(-U_i)}{2}, \qquad \text{with i.i.d. } U_i \sim \mathsf{N}(0, 1). \qquad (21.19)$$

Because the $\mathsf{N}(0, 1)$ distribution is symmetric about the origin, rather than about $\frac{1}{2}$, the antithetic estimate uses $-U_i$, rather than $1 - U_i$. Of course, $-U_i$ is also an $\mathsf{N}(0, 1)$ random variable.

The above analysis that gave us (21.16) can then be repeated to give us

$$\text{var}\left(\frac{f(U_i) + f(-U_i)}{2}\right) \leq \tfrac{1}{2}\text{var}\left(f(U_i)\right) \qquad (21.20)$$

when f is monotonic.

Computational example Here we show the antithetic variate trick in use with $\mathsf{N}(0, 1)$ samples. We take (21.17) with $f(x) = (1/\sqrt{e})e^x$, so that $\mathbb{E}(f(U)) = 1$ (see Exercise 15.3). (A similar computation was done in Chapter 15 for standard Monte Carlo. We now scale by $1/\sqrt{e}$ so that the confidence intervals are easier to assimilate.) Table 21.2 shows the 95% confidence intervals for (21.18) and (21.19). As in the previous example, we took $M = 10^2, 10^3, 10^4, 10^5$, and used the same random number samples for the two methods. The antithetic version gives almost twice as much accuracy. ◇

Table 21.2. *Ninety-five per cent confidence intervals for (21.18) and (21.19) on problem (21.17) with $f(x) = (1/\sqrt{e})e^x$, plus ratios of their widths*

M	Standard	Antithetic	Ratio of widths
10^2	[0.8247, 1.2819]	[0.9518, 1.6767]	0.6
10^3	[0.9713, 1.1574]	[1.0166, 1.1244]	1.7
10^4	[0.9647, 1.0137]	[0.9945, 1.0243]	1.6
10^5	[0.9953, 1.0115]	[0.9955, 1.0046]	1.8

21.7 Multivariate case

The antithetic variates idea extends readily to the case where f is a function of more than one random variable. For example, suppose we wish to approximate

$$I = \mathbb{E}(f(U, V, W)), \qquad \text{where } U, V, W \text{ are i.i.d.} \sim \mathsf{N}(0, 1).$$

The standard Monte Carlo estimate is

$$I_M = \frac{1}{M} \sum_{i=1}^{M} f(U_i, V_i, W_i), \qquad \text{with } U_i, V_i, W_i \text{ i.i.d.} \sim \mathsf{N}(0, 1),$$

and the antithetic version is

$$\widehat{I_M} = \frac{1}{M} \sum_{i=1}^{M} \frac{f(U_i, V_i, W_i) + f(-U_i, -V_i, -W_i)}{2},$$

$$\text{with } U_i, V_i, W_i \text{ i.i.d.} \sim \mathsf{N}(0, 1).$$

An extension of the above analysis shows that benefits accrue when f is monotonic in each of the arguments.

21.8 Antithetic variates in option valuation

The application that we have in mind is, of course, Monte Carlo estimation of path-dependent exotic options. In this case we discretize the time interval $[0, T]$ and compute risk-neutral asset prices at $\{t_i\}_{i=1}^{N}$, with $t_i = i \Delta t$, $N \Delta t = T$. We know that on each increment the price update uses an $\mathsf{N}(0, 1)$ random variable Z_j coming from the i.i.d. sequence $\{Z_0, Z_1, \ldots, Z_{N-1}\}$ according to (19.7). We wish to compute the expected value of some payoff function. We are therefore looking for the expected value of a function of the N i.i.d. $\mathsf{N}(0, 1)$ random variables $\{Z_0, Z_1, \ldots, Z_{N-1}\}$. The antithetic variates technique is to take the average payoff from one path with samples $\{Z_0, Z_1, \ldots, Z_{N-1}\}$ and another path with

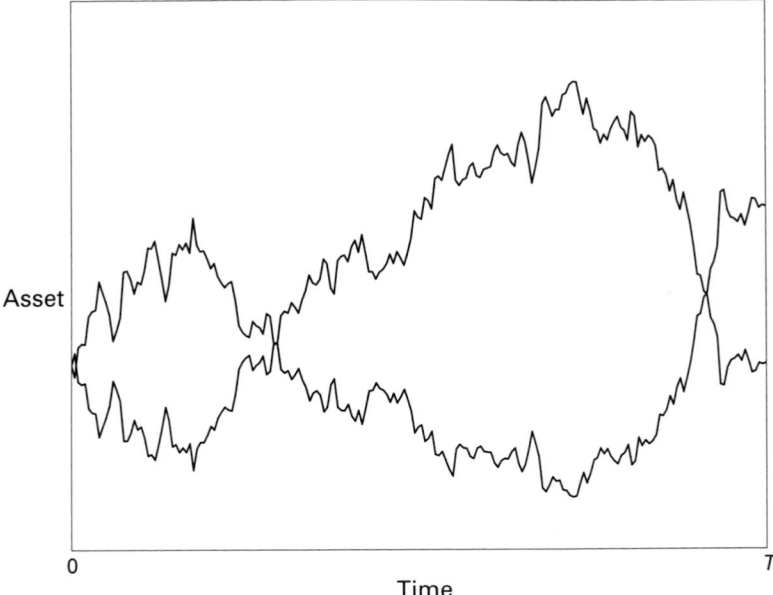

Fig. 21.1. A pair of discrete asset paths computed using antithetic variates. The payoff from both paths is averaged in order to give a single sample.

samples $\{-Z_0, -Z_1, \ldots, -Z_{N-1}\}$. Where one path zig-zags, the other path zag-zigs. Figure 21.1 illustrates such a pair of paths.

Computational example We value an up-and-in call option with $S_0 = 5$, $E = 6$, $r = 0.05$, $\sigma = 0.3$ and $T = 1$, using a timestep $\Delta t = 10^{-4}$, so $N = 10^4$. We take $B = 8$ for the barrier level. Recall from Section 19.2 that

- the payoff is zero if the asset never attained the price B, that is, if $\max_{[0,T]} S(t) < B$,
- the payoff is equal to the European call value $\max(S(T) - E, 0)$ if the asset attained the price B, that is, if $\max_{[0,T]} S(t) \geq B$.

Using the ideas from Section 19.6, a basic Monte Carlo strategy can be summarized as follows:

```
for i = 1 to M
    for j = 0 to N − 1
        compute an N(0, 1) sample ξⱼ
        set S_{j+1} = S_j e^{(r−½σ²)Δt+σ√Δt ξ_j}
    end
    set Sᵢᵐᵃˣ = max_{0≤j≤N} S_j
    if Sᵢᵐᵃˣ > B set Vᵢ = e^{−rT} max(S_N − E, 0), otherwise Vᵢ = 0
```

```
end
```
$$\text{set } a_M = \left(\frac{1}{M}\right) \sum_{i=1}^{M} V_i$$
$$\text{set } b_M^2 = \left(\frac{1}{M-1}\right) \sum_{i=1}^{M} (V_i - a_M)^2$$

This gives an approximate option price a_M and an approximate 95% confidence interval (15.5).

The corresponding antithetic variate version is

```
for i = 1 to M
        for j = 0 to N − 1
```
$$\text{compute an N}(0, 1) \text{ sample } \xi_j$$
$$\text{set } S_{j+1} = S_j e^{(r-\frac{1}{2}\sigma^2)\Delta t + \sigma\sqrt{\Delta t}\xi_j}$$
$$\text{set } \overline{S}_{j+1} = \overline{S}_j e^{(r-\frac{1}{2}\sigma^2)\Delta t - \sigma\sqrt{\Delta t}\xi_j}$$
```
        end
```
$$\text{set } S_i^{\max} = \max_{0 \le j \le N} S_j$$
$$\text{set } \overline{S}_i^{\max} = \max_{0 \le j \le N} \overline{S}_j$$
$$\text{if } S_i^{\max} > B \text{ set } V_i = e^{-rT} \max(S_N - E, 0), \text{ otherwise } V_i = 0$$
$$\text{if } \overline{S}_i^{\max} > B \text{ set } \overline{V}_i = e^{-rT} \max(\overline{S}_N - E, 0), \text{ otherwise } \overline{V}_i = 0$$
$$\text{set } \widehat{V}_i = \frac{1}{2}(V_i + \overline{V}_i)$$
```
end
```
$$\text{set } a_M = \frac{1}{M} \sum_{i=1}^{M} \widehat{V}_i$$
$$\text{set } b_M^2 = \frac{1}{M-1} \sum_{i=1}^{M} (\widehat{V}_i - a_M)^2$$

Table 21.3 shows the 95% confidence intervals, and the ratios of their widths, for $M = 10^2$, 10^3, 10^4, 10^5. We see that using antithetic variates shrinks the confidence intervals by a factor of around 1.5. As mentioned in Section 19.6, the overall accuracy of the process depends not only on the error in the Monte Carlo approximation to the mean, but also on the error arising from the time discretization – we take the maximum over a discrete set of points rather than over a continuous time interval. In this experiment we found that using smaller Δt values did not significantly change the computed results, so the sampling error is dominant.   ◇

Table 21.3. *Ninety-five per cent confidence intervals, plus ratios of their widths, for standard and antithetic Monte Carlo on an up-and-in call*

M	Standard	Antithetic	Ratio of widths
10^2	[0.0878, 0.3219]	[0.1239, 0.3061]	1.3
10^3	[0.2285, 0.3333]	[0.2238, 0.2936]	1.5
10^4	[0.2443, 0.2764]	[0.2370, 0.2580]	1.5
10^5	[0.2359, 0.2458]	[0.2373, 0.2440]	1.5

21.9 Notes and references

The texts (Hammersley and Handscombe, 1964; Madras, 2002; Ripley, 1987) that we mentioned in Chapter 15 are good sources of general information about antithetic variates, and (Boyle *et al.*, 1997; Boyle, 1977; Clewlow and Strickland, 1998; Jäckel, 2002) look at practical issues for option valuation.

EXERCISES

21.1. ⋆ Show that (21.1) and (21.2) are equivalent and hence conclude that if X and Y are independent then $\mathsf{cov}(X, Y) = 0$.

21.2. ⋆ Show that $I = 2$ in (21.3) and confirm (21.5).

21.3. ⋆ Establish the identity (21.7). [Hints: make use of (3.6) and (3.10) in (21.1).]

21.4. ⋆⋆ Use your favourite scientific computation package to confirm that $\mathsf{var}(\widehat{Y}_i) \approx 0.001073$ in (21.10). (For example, a suitable approximation to the integral $\int_0^1 e^{\sqrt{x}+\sqrt{1-x}}\, dx$ in (21.10) can be obtained from `>> quadl('exp(sqrt(x) + sqrt(1-x))',0,1,1e-9)` in MATLAB.)

21.5. ⋆⋆ Prove the statement involving (21.15).

21.6. ⋆⋆ Consider the case where f is a monotonic increasing function that is extremely expensive to evaluate on a computer – so much so that the cost of a sample from a pseudo-random number generator is negligible by comparison. Can we still argue that the antithetic variate estimate (21.13) is at least as efficient as the standard one, (21.12)?

21.7. ⋆⋆ Show that the antithetic estimators (21.13) and (21.19) are exact in the case where f is linear, that is, $f(x) = \alpha x + \beta$, for $\alpha, \beta \in \mathbb{R}$. What can you say about the corresponding confidence intervals?

21.8. ⋆⋆⋆ Find a simple example where antithetic variates are *less efficient* than standard Monte Carlo.

21.10 Program of Chapter 21 and walkthrough

In ch21, listed in Figure 21.2, we value an up-and-out call option. We use the same parameters as for ch19, so we know that the Black–Scholes value is 0.1857.

The first part of the for loop implements standard Monte Carlo, as in ch19. We then compute the payoffs with a negated version of the pseudo-random numbers in samples. The ith entry of the array Vanti thus contains the average of the payoffs for the ith asset path and its antithetic twin.

Running ch21 gives conf = [0.1763, 0.1937] for the Monte Carlo confidence interval. This is identical to the interval produced by ch19, because by setting the random number generator to the same state with randn('state',100), we are using exactly the same samples. The antithetic version gives confanti = [0.1807, 0.1921], which is roughly 1.5 times as small as the standard Monte Carlo confidence interval.

```
%CH21        Program for Chapter 21
%
% Up-and-out call option
% Uses Monte Carlo with antithetic variates

randn('state',100)

%%%%%% Problem and method parameters %%%%%%%%%
S = 5;  E = 6;  sigma = 0.25;  r = 0.05;  T = 1; B = 9;
Dt = 1e-3; N = T/Dt; M = 1e4;
%%%%%%%%%%%%%%%%%%%%%%%%%%%%%%%%%%%%%%%%%%%%

V = zeros(M,1);
Vanti = zeros(M,1);
for i = 1:M
  samples = randn(N,1);

  % standard Monte Carlo
  Svals = S*cumprod(exp((r-0.5*sigma^2)*Dt+sigma*sqrt(Dt)*samples));
  Smax = max(Svals);
  if Smax < B
    V(i) = exp(-r*T)*max(Svals(end)-E,0);
  end

  % antithetic path
  Svals2 = S*cumprod(exp((r-0.5*sigma^2)*Dt-sigma*sqrt(Dt)*samples));
  Smax2 = max(Svals2);
  V2 = 0;
  if Smax2 < B
    V2 = exp(-r*T)*max(Svals2(end)-E,0);
  end
  Vanti(i) = 0.5*(V(i) + V2);

end
aM = mean(V); bM = std(V);
conf = [aM - 1.96*bM/sqrt(M), aM + 1.96*bM/sqrt(M)]

aManti = mean(Vanti); bManti = std(Vanti);
confanti = [aManti - 1.96*bManti/sqrt(M), aManti + 1.96*bManti/sqrt(M)]
```

Fig. 21.2.   Program of Chapter 21: ch21.m.

PROGRAMMING EXERCISES

P21.1. Alter `ch21` to the case of a different exotic option.

P21.2. Type `help cov` to learn about MATLAB's covariance function, and apply it to the examples studied in this chapter.

Quotes

Monte Carlo simulation will continue to gain appeal
as financial instruments become more complex, workstations become faster,
and simulation software is adopted by more users.
The use of variance reduction techniques
along with the greater power of today's workstations
can help to reduce the execution time required for achieving acceptable precision
to the point that simulation can be used by financial traders to value derivatives in real time.
> JOHN CHARNES, 'Sharper estimates of derivative values', *Financial Engineering News*, June/July 2002, Issue No. 26

Even statisticians often fail to treat simulations seriously as experiments.
> BRIAN D. RIPLEY (Ripley, 1987)

It's not always easy to tell the difference between understanding
and brute force computation.
> ROGER PENROSE, source www.apmaths.uwo.ca/ rcorless/

22

Monte Carlo Part III: variance reduction by control variates

22.1 Motivation

We saw in the previous chapter that the antithetic variates idea relies upon finding samples that are anticorrelated with the original random variable. In contrast, the technique discussed here relies upon finding samples that have some general *known* correlation. This control variate approach is less generic than antithetic variates, as it requires some knowledge about the underlying random variables in the simulations. However, when it works it can be very powerful.

22.2 Control variates

Given that we wish to estimate $\mathbb{E}(X)$, suppose we can somehow find another random variable, Y, that is 'close' to X with known mean $\mathbb{E}(Y)$. Then the random variable

$$Z = X + \mathbb{E}(Y) - Y \tag{22.1}$$

satisfies $\mathbb{E}(Z) = \mathbb{E}(X) + \mathbb{E}(Y) - \mathbb{E}(Y) = \mathbb{E}(X)$, and hence we could apply Monte Carlo to Z instead of X. In this context, Y is called the *control variate*.

Since adding a constant to a random variable does not change its variance (Exercise 3.6), we see immediately from (22.1) that

$$\mathsf{var}(Z) = \mathsf{var}(X - Y).$$

Hence, to get some benefit from this approach, we would like $X - Y$ to have a smaller variance than X. This is what we mean above by 'close'. Note, however, that it may be more expensive to sample Z than X. If $\mathsf{var}(Z) = R_1 \mathsf{var}(X)$ for

229

some $R_1 < 1$ and the cost of sampling Z is R_2 times that of sampling X, then we get an overall gain in efficiency if $R_1 R_2 < 1$, see Exercise 22.1.

We may generalize (22.1) to the case of

$$Z_\theta = X + \theta \left(\mathbb{E}(Y) - Y \right), \tag{22.2}$$

for any $\theta \in \mathbb{R}$. Note that we still have $\mathbb{E}(Z_\theta) = \mathbb{E}(X)$, so we may apply Monte Carlo to Z_θ. In this case

$$\mathsf{var}(Z_\theta) = \mathsf{var}(X - \theta Y) = \mathsf{var}(X) - 2\theta \, \mathsf{cov}(X, Y) + \theta^2 \mathsf{var}(Y).$$

As θ varies, the value of θ that minimizes this quadratic is given by

$$\theta_{\min} := \frac{\mathsf{cov}(X, Y)}{\mathsf{var}(Y)}. \tag{22.3}$$

Further, we can show that $\mathsf{var}(Z_\theta) < \mathsf{var}(X)$ if and only if θ lies between 0 and $2\theta_{\min}$, see Exercise 22.2.

Of course, on a general problem we typically do not know $\mathsf{cov}(X, Y)$ and hence cannot find $\theta_{\min}$. However, it is possible to estimate $\mathsf{cov}(X, Y)$, and hence $\theta_{\min}$, during a Monte Carlo simulation.

The name 'control variate' comes from the fact that the $\mathbb{E}(Y) - Y$ term controls the Monte Carlo process. Suppose the covariance is positive, that is, $\mathsf{cov}(X, Y) := \mathbb{E}((X - \mathbb{E}(X)) (Y - \mathbb{E}(Y))) > 0$ and $\theta > 0$. In this case, when X is larger than average $(X > \mathbb{E}(X))$ we would also expect Y to be larger than average $(Y > \mathbb{E}(Y))$. Generally, adding the negative amount $\theta \left(\mathbb{E}(Y) - Y \right)$ helps to correct the overestimate of $\mathbb{E}(X)$ from that sample of X. Similarly when X is smaller than average $(X < \mathbb{E}(X))$ we would also expect Y to be smaller than average $(Y < \mathbb{E}(Y))$ and adding the positive amount $\theta \left(\mathbb{E}(Y) - Y \right)$ helps to correct the underestimate. A similar argument applies when $\mathsf{cov}(X, Y) < 0$ and $\theta < 0$.

Computational example We return to the example from the previous chapter of computing $I = \mathbb{E}(e^{\sqrt{U}})$, where $U \sim \mathsf{U}(0, 1)$. For illustration, we take e^U as our control variate, and use the fact that $\mathbb{E}(e^U) = \int_0^1 e^x \, dx = e - 1$. Since $e^{\sqrt{U}}$ and e^U are close over $[0, 1]$, we will try the simple $\theta = 1$ version. Thus the control variate Monte Carlo algorithm applies to $Z = e^{\sqrt{U}} + e - 1 - e^U$. Table 22.1 shows the 95% confidence intervals for the standard and control variate algorithms, and also the ratios of confidence interval widths. We did four tests, covering $M = 10^2, 10^3, 10^4, 10^5$, and used the same random number samples for the two methods. (Note that the confidence intervals for standard Monte Carlo are identical to those in Table 21.1, as we started the random number generator at the same point.) We see that the control variate version has confidence intervals that are just over 4 times smaller. Separate computations confirm that

Table 22.1. *Ninety-five per cent confidence intervals with standard and control variate algorithm (22.1) for* $\mathbb{E}(e^{\sqrt{U}})$, *plus ratios of their widths*

M	Standard	Control variate	Ratio of widths
10^2	[1.8841, 2.0752]	[1.9601, 2.0031]	4.4
10^3	[1.9538, 2.0087]	[1.9951, 2.0084]	4.1
10^4	[1.9890, 2.0062]	[1.9994, 2.0036]	4.1
10^5	[1.9969, 2.0023]	[1.9993, 2.0006]	4.1

Table 22.2. *Ninety-five per cent confidence intervals with standard and control variate algorithm (22.2) for* $\mathbb{E}(e^{\sqrt{U}})$, *plus ratios of their widths*

M	Standard	θ-Control variate	θ	Ratio of widths
10^2	[1.8841, 2.0752]	[1.9623, 2.0004]	0.89	5.0
10^3	[1.9538, 2.0087]	[1.9937, 2.0048]	0.88	4.9
10^4	[1.9890, 2.0062]	[1.9993, 2.0027]	0.88	5.0
10^5	[1.9969, 2.0023]	[1.9994, 2.0005]	0.88	5.0

$\mathsf{var}(e^{\sqrt{U}} - e^{U})$ is about 17 times smaller than $\mathsf{var}(e^{\sqrt{U}})$. Next, we tried the more general version based on (22.2). Here, we initially used the $\mathsf{U}(0, 1)$ samples from the random number generator to estimate $\mathsf{cov}(X, Y)$ and $\mathsf{var}(Y)$, and hence estimate $\theta_{\min}$ in (22.3). The samples were then re-used for the Monte Carlo estimate of (22.2) with this θ value. Table 22.2 gives the results, including the θ values that arose. We see that the optimal θ estimates are close to 1, and the extra work has only slightly improved the confidence interval widths. ◇

22.3 Control variates in option valuation

The control variate idea can be used on path-dependent options where there is no known analytical expression for the option value, but there **is** an expression for a similar option. The classic example is an *arithmetic average price Asian* option, where the average is taken over a pre-set collection of discrete times $\{t_i\}_{i=1}^{n}$. As described in Section 19.4, the payoff for the arithmetic average price Asian call option is

$$\max\left[\frac{1}{n}\sum_{i=1}^{n} S(t_i) - E, 0\right], \tag{22.4}$$

whereas the corresponding *geometric average price Asian* option has payoff

$$\max\left[\left(\prod_{i=1}^{n} S(t_i)\right)^{1/n} - E, 0\right]. \tag{22.5}$$

We see that (22.5) differs from (22.4) only in that the arithmetic average has been replaced by a geometric average. If the discrete times are equally spaced, $t_i = i\,\Delta t$, with $\Delta t = T/n$, then Exercise 19.6 shows that there is an exact formula for the geometric average option. However, for the arithmetic average version there is no known explicit formula.

It is reasonable to expect the arithmetic and geometric versions to be well correlated – typically, paths where one option has a large/small payoff should also be paths where the other option has a large/small payoff. Because we have the exact expression (19.10) for the value (that is, the expected payoff under risk neutrality) of the geometric version, we may use this option as a control variate when valuing the arithmetic version.

Computational example We now use Monte Carlo to value the arithmetic average price Asian option described above. We take $S_0 = 5$, $E = 6$, $r = 0.05$, $\sigma = 0.3$ and $T = 1$, and discrete time points $\Delta t, 2\Delta t, \ldots, n\Delta t$, where $n = 100$, so $\Delta t = 10^{-2}$. Since we are not interested in the asset prices at any other times, we used Δt as the timestep in the algorithm and computed risk-neutral asset prices $S(\Delta t), S(2\Delta t), \ldots, S(N\Delta t)$. Table 22.3 shows the 95% confidence intervals for standard Monte Carlo and for the alternative that uses the geometric average price Asian option as a control variate in the basic formulation (22.1). We used $M = 10^2, 10^3, 10^4, 10^5$ samples. We see that the control variate improves accuracy by a factor of around eight. In this case, sampling the control variate involves relatively little extra work, so the gain in efficiency is significant.   ◇

22.4 Notes and references

The references (Hammersley and Handscombe, 1964; Madras, 2002; Ripley, 1987) deal with the use of control variates in general, and (Boyle *et al.*, 1997; Boyle, 1977; Clewlow and Strickland, 1998; Jäckel, 2002) apply specifically to finance. The review paper (Boyle *et al.*, 1997) also discusses a number of other variance reduction techniques.

Because of the representation (3.8), any algorithm for approximating an expected value may be thought of as a *quadrature* method, that is, a method for approximating integrals. Quadrature has a long and distinguished history in numerical analysis, and many methods have been developed. Monte Carlo

Table 22.3. *Ninety-five per cent confidence intervals with*
standard and $\theta = 1$ *control variate algorithm (22.1) for an*
Asian option, plus ratios of their widths

M	Standard	Control Variate	Ratio of widths
10^2	[0.0283, 0.1161]	[0.0885, 0.1010]	7.1
10^3	[0.0823, 0.1207]	[0.0947, 0.0990]	8.9
10^4	[0.0911, 0.1035]	[0.0965, 0.0981]	8.2
10^5	[0.0968, 0.1007]	[0.0973, 0.0978]	8.2

simulations for path-dependent options, where asset paths are computed at points $\Delta t, 2\Delta t, \ldots, N\Delta t = T$, correspond to N-dimensional integrals. In this context, although Monte Carlo is one of the few viable techniques, current research indicates that algorithms based on so-called *low discrepancy sequences* can be more efficient. *Quasi Monte Carlo* methods, which combine the efficiency of low discrepancy sequences with the confidence interval information from Monte Carlo, have also been developed recently. The texts (Hull, 2000; Jäckel, 2002; Kwok, 1998) and the survey (Boyle *et al.*, 1997) give pointers to recent literature.

Both variance reduction and hedging share the aim of making a random variable more predictable, and this connection can be exploited in practice, see (Clewlow and Strickland, 1998), for example.

EXERCISES

22.1. ⋆⋆ Confirm that if $\mathsf{var}(Z) = R_1 \mathsf{var}(X)$ for some $R_1 < 1$ and the cost of sampling Z is R_2 times that of sampling X, then we get an overall gain in efficiency from applying Monte Carlo to (22.1) if $R_1 R_2 < 1$.

22.2. ⋆⋆ Show that $\mathsf{var}(Z_\theta) < \mathsf{var}(X)$ if and only if θ lies between 0 and $2\theta_{\min}$, where $\theta_{\min}$ is defined in (22.3). (Note that $\theta_{\min}$ may be negative.)

22.3. ⋆ ⋆ ⋆ (This exercise relates to Section 15.4, but fits in with the general theme of variance reduction.) Suppose that a random variable V depends on some deterministic parameter, p, and we wish to compute

$$\frac{\mathbb{E}\left(V(p+h)\right) - \mathbb{E}\left(V(p)\right)}{h}$$

for some small increment h. Consider the following Monte Carlo approaches:

Method 1

(a) apply Monte Carlo to give an approximation $A \approx \mathbb{E}\,(V\,(p+h))$,
(b) apply Monte Carlo to give an approximation $B \approx \mathbb{E}\,(V\,(p))$ (using a different pseudo-random number sequence from that in (a)),
(c) compute $(A - B)/h$ as the overall approximation.

Method 2

(a) apply Monte Carlo to give an approximation $C \approx \mathbb{E}(V\,(p+h) - V\,(p))$,
(b) compute C/h as the overall approximation.

Using (21.7), explain why Method 2 is likely to be more successful than Method 1.

22.5 Program of Chapter 22 and walkthrough

In ch22, listed in Figure 22.1, we do a control variate computation of the type reported in Table 22.3. Our task is to value an arithmetic average price Asian option using the geometric average price Asian, which has a Black–Scholes formula, as control variate. After initializing the parameters, we evaluate the formula (19.10) for the geometric version. Next, we compute an M by L array Spath, whose ith row represents the ith asset path at times $0, \Delta t, 2\Delta t, \ldots, T$. The standard Monte Carlo method is then applied. The command mean(Spath,2) evaluates the sample mean over the second index; this returns an M by 1 array whose ith entry is the sample mean over the ith row of Spath. The quantity exp(-r*T)*max(arithave-E,0) then represents the array whose ith entry is the payoff of the arithmetic average price Asian option from the ith path. The Monte Carlo confidence interval for these payoff samples turned out to be confmc = [0.2479,0.2631].

For the geometric average price Asian control variate, we must evaluate the quantity

$$\left(\prod_{i=1}^{N} S(t_i) \right)^{1/N}.$$

To eliminate the possibility of under/overflow in the evaluation of the product $\prod_{i=1}^{N} S(t_i)$ it is prudent to implement the equivalent form

$$\exp\left(\frac{1}{N} \sum_{i=1}^{N} \log\,(S(t_i)) \right).$$

The variable geoave gives the pathwise geometric average and Pgeo then stores the payoffs – these are our control variate samples. The array Z thus contains samples corresponding to Z in (22.1). The resulting confidence interval is confcv = [0.2576, 0.2584]. This is nearly 20 times smaller than confmc.

PROGRAMMING EXERCISES

P22.1. Alter ch22 so that the θ version (22.2) is used.

P22.2. Test whether it is worthwhile to combine the antithetic and control variates techniques.

```
%CH22      Program for Chapter 22
%
% Monte Carlo on an arithmetic average price Asian option
% using a geometric average price Asian as control variate

randn('state',100)

%%%%% Problem and method parameters %%%%%%%%%
S = 4; E = 4; sigma = 0.25; r = 0.03; T = 1;
Dt = 1e-2; N = T/Dt; M = 1e4;
%%%%%%%%%%%%%%%%%%%%%%%%%%%%%%%%%%%%%%%%%

%%%%%%%%% Geom Asian exact mean  %%%%%%%%%%%%%%
sigsqT= sigma^2*T*(N+1)*(2*N+1)/(6*N*N);
muT = 0.5*sigsqT + (r - 0.5*sigma^2)*T*(N+1)/(2*N);

d1 = (log(S/E) + (muT + 0.5*sigsqT))/(sqrt(sigsqT));
d2 = d1 - sqrt(sigsqT);

N1 = 0.5*(1+erf(d1/sqrt(2)));
N2 = 0.5*(1+erf(d2/sqrt(2)));

geo =  exp(-r*T)*( S*exp(muT)*N1 - E*N2 );
%%%%%%%%%%%%%%%%%%%%%%%%%%%%%%%%%%%%%%%

Spath = S*cumprod(exp((r-0.5*sigma^2)*Dt+sigma*sqrt(Dt)*randn(M,N)),2);

% Standard Monte Carlo
arithave = mean(Spath,2);
Parith = exp(-r*T)*max(arithave-E,0);  % payoffs
Pmean = mean(Parith);
Pstd = std(Parith);
confmc = [Pmean-1.96*Pstd/sqrt(M), Pmean+1.96*Pstd/sqrt(M)]

% Control variate
geoave = exp((1/N)*sum(log(Spath),2));
Pgeo = exp(-r*T)*max(geoave-E,0);     % geo payoffs
Z = Parith + geo - Pgeo;              % control variate version
Zmean = mean(Z);
Zstd = std(Z);
confcv = [Zmean-1.96*Zstd/sqrt(M), Zmean+1.96*Zstd/sqrt(M)]
```

Fig. 22.1.   Program of Chapter 22: ch22.m.

Quotes

Simulation has a colourful language, and variance reduction techniques,
especially clever ones, are often known as *swindles*.
Presumably it is nature that is being swindled,
but she frequently gets her own back.
Variance reduction swindles quite frequently do not work,
especially when more than one idea is tried simultaneously.

BRIAN D. RIPLEY (Ripley, 1987)

The antithetic method is easy to implement,
but often leads to only modest error reductions.
...The control variate technique can lead to very substantial error reductions,
but its effectiveness hinges on finding a good control for each problem.

PHELIM BOYLE, MARK BROADIE AND PAUL GLASSERMAN (Boyle *et al.*, 1997)

Spock: Random chance seems to have operated in our favor.
McCoy: In plain non-Vulcan English, we've been lucky!
Spock: I believe I said that, doctor.

From STAR TREK, source http://us.imdb.com/Quotes?0060028

Never let the continuous progress of CPU speeds
and processing power
be an excuse for ill-thought-out algorithm design.

PETER JÄCKEL (Jäckel, 2002)

23

Finite difference methods

23.1 Motivation

In Chapter 8 we obtained the Black–Scholes formula for a European call option by first deriving the PDE (8.15)–(8.18) and then displaying its analytical solution (8.19). Chapters 18 and 19 showed that the values of other options may also be characterized via the solutions to PDEs. In general, the PDEs that arise for valuing exotic options cannot be solved analytically. However, it is possible to compute approximate solutions. This chapter introduces finite difference methods, which represent the most popular computational approach. We have already come across the underlying idea, that of discretization, a number of times in this book. Here, we develop three widely used finite difference schemes for the basic heat equation and discuss their key properties. The next chapter focuses on the use of finite difference technology for option valuation.

23.2 Finite difference operators

Given a smooth function $y : \mathbb{R} \to \mathbb{R}$, we know from the definition of a derivative that, for small h,

$$\frac{y(x + h) - y(x)}{h} \approx \frac{dy}{dx}(x).$$

Table 23.1. *Difference operators*

Operator	Symbol	Definition	Taylor series
Forward difference	Δ	$y_{m+1} - y_m$	$hy' + \frac{1}{2}h^2 y'' + \ldots$
Backward difference	∇	$y_m - y_{m-1}$	$hy' - \frac{1}{2}h^2 y'' + \ldots$
Half central difference	δ	$y_{m+\frac{1}{2}} - y_{m-\frac{1}{2}}$	$hy' - \frac{1}{24}h^2 y'' + \ldots$
Full central difference	Δ_0	$\frac{1}{2}(y_{m+1} - y_{m-1})$	$hy' + \frac{1}{6}h^3 y''' + \ldots$
Second order central difference	δ^2	$y_{m+1} - 2y_m + y_{m-1}$	$h^2 y'' - \frac{1}{12}h^4 y'''' + \ldots$
Shift	E	y_{m+1}	$y + hy' + \ldots$
Average	μ	$\frac{1}{2}\left(y_{m+\frac{1}{2}} + y_{m-\frac{1}{2}}\right)$	$y + \frac{1}{8}h^2 y'' + \ldots$

If we let y_m denote the value $y(mh)$ then this may be written

$$y_{m+1} - y_m \approx h\frac{dy}{dx}(mh). \tag{23.1}$$

To ease the notation, we will use a prime, $'$, to denote a derivative, so y' means dy/dx and y'' means $d^2 y/dx^2$, and assume that functions are evaluated at $x = mh$ unless otherwise stated. With the aid of a simple Taylor series expansion, we may extend (23.1) to

$$y_{m+1} - y_m = hy' + \frac{1}{2}h^2 y'' + \cdots$$

The quantity $y_{m+1} - y_m$ is known as a *forward difference* and the associated *forward difference operator*, Δ, is defined by

$$\Delta y_m = y_{m+1} - y_m.$$

(This is, of course, not to be confused with the delta of an option. In this chapter Δ will exclusively denote the forward difference operator.) In Table 23.1 we define a number of finite difference operators. These operators, which act on grid values $y_m = y(mh)$, form the main building blocks of finite difference methods. The table also gives the first two terms in the corresponding Taylor series expansions; Exercise 23.2 asks you to verify them. Two of those definitions involve 'half-way' values, $y_{m\pm\frac{1}{2}} = y((m \pm \frac{1}{2})h)$.

23.3 Heat equation

We will focus on a simple mathematical problem. Our task is to find a function of two variables, $u(x, t)$, that satisfies the PDE

$$\frac{\partial u}{\partial t} = \frac{\partial^2 u}{\partial x^2}, \qquad \text{for} \qquad 0 \le x \le L \qquad \text{and} \qquad 0 \le t \le T, \tag{23.2}$$

subject to the initial condition

$$u(x, 0) = g(x) \qquad (23.3)$$

and the boundary conditions

$$u(0, t) = a(t) \qquad \text{and} \qquad u(L, t) = b(t). \qquad (23.4)$$

The PDE (23.2) is known as the *heat equation*, because $u(x, t)$ describes the temperature at time t of the point x on a thin metal bar with initial temperature profile (23.3) and with endpoints heated according to (23.4). There are two good reasons for focusing on this PDE.

 (i) It allows us to develop the ideas behind finite difference methods in a simple setting.
(ii) The basic Black–Scholes PDE (8.15) can be translated into an equation of this form; Section 23.9 gives references. In Section 24.4 we will go part of the way towards that translation.

We will follow the usual convention of regarding x as space and t as time. (In the next chapter, however, x will correspond to asset price.)

In the case $L = \pi$ with

$$g(x) = \sin(x), \quad a(t) = b(t) = 0, \qquad (23.5)$$

it is easy to verify that

$$u(x, t) = e^{-t} \sin(x) \qquad (23.6)$$

solves (23.2), (23.3) and (23.4). In Figure 23.1 we plot the solution (23.6). This will be used for reference when we derive finite difference methods.

23.4 Discretization

In computing an approximate solution to the PDE (23.2), (23.3) and (23.4), our first step is to *discretize*. We have already used the idea of *discretization* in a number of contexts:

• development of an asset price model in Chapter 6,
• derivation of the binomial method in Chapter 16,
• extension of Monte Carlo to path-dependent option valuation in Chapter 19.

The plan is to compute approximations to the PDE solution only at a finite set of points. We divide the space axis into $N_x + 1$ equally spaced points $\{jh\}_{j=0}^{N_x}$, where $h = L/N_x$, and the time axis into $N_t + 1$ equally spaced points $\{ik\}_{i=0}^{N_t}$, where $k = T/N_t$. The points (jh, ik) form what is called the *grid*, or

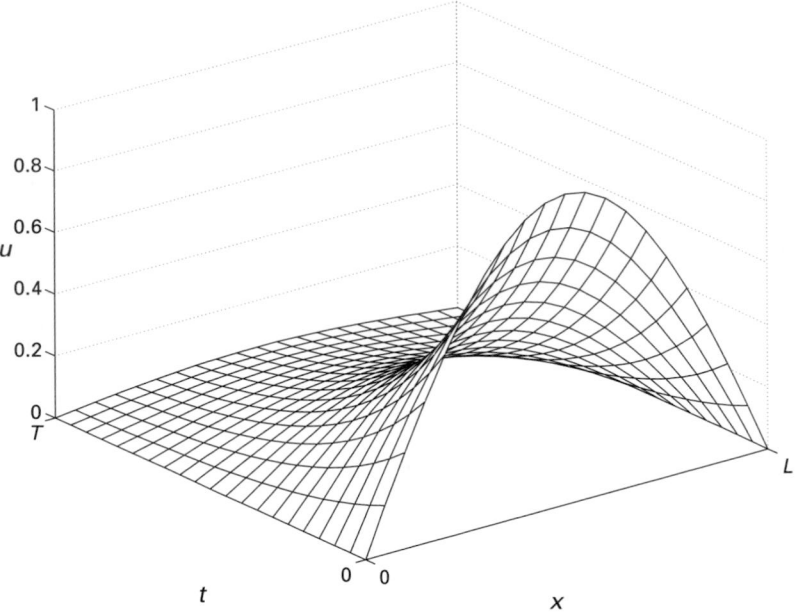

Fig. 23.1.   Heat equation solution $u(x, t)$ for (23.2), (23.3) and (23.4) with initial and boundary conditions (23.5).

the *mesh*. We seek values U^i_j that approximate the solution on the grid,

$$U^i_j \approx u(jh, ik), \qquad 0 \le j \le N_x \quad \text{and} \quad 0 \le i \le N_t.$$

This notation is consistent with that in Chapter 16 in the sense that a superscript is used to denote the time level. Figure 23.2 illustrates the grid. The open circles indicate grid points where the solution is not yet known. Points where the initial condition (23.3) and boundary conditions (23.4) can be used to determine the solution are marked with filled circles. Hence, our task is to find numbers to put into the points marked ○. We will do this by using finite difference operators to form equations that the grid values U^i_j must satisfy.

23.5 FTCS and BTCS

The key step in deriving finite difference methods is to replace differential operators with finite difference operators. Our problem domain involves two independent variables, $0 \le x \le L$ and $0 \le t \le T$, and hence we must distinguish between difference operators in the x- and t-directions. We do this with a subscript, so, for example,

$$\Delta_t U^i_j = U^{i+1}_j - U^i_j \qquad \text{and} \qquad \Delta_x U^i_j = U^i_{j+1} - U^i_j.$$

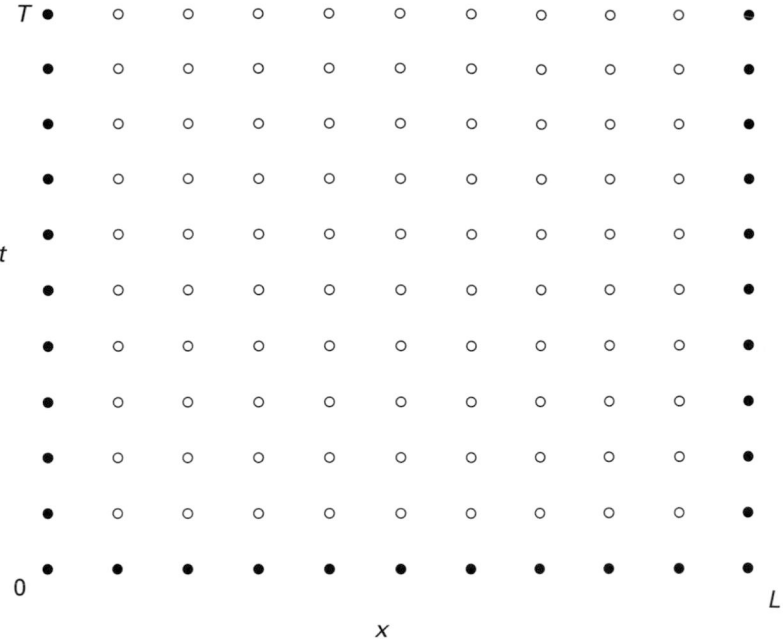

Fig. 23.2. Finite difference grid $\{jh, ik\}_{j=0,i=0}^{N_x, N_t}$. Points are spaced at a distance of h apart in the x-direction and k apart in the t-direction.

A simple method for the heat equation (23.2) involves approximating the time derivative $\partial/\partial t$ by the scaled forward difference in time, $k^{-1}\Delta_t$, and the second order space derivative $\partial^2/\partial x^2$ by the scaled second order central difference in space, $h^{-2}\delta_x^2$. This gives the equation

$$k^{-1}\Delta_t U_j^i - h^{-2}\delta_x^2 U_j^i = 0,$$

which may be expanded as

$$\frac{U_j^{i+1} - U_j^i}{k} - \frac{U_{j+1}^i - 2U_j^i + U_{j-1}^i}{h^2} = 0.$$

A more revealing re-write is

$$U_j^{i+1} = \nu U_{j+1}^i + (1 - 2\nu)U_j^i + \nu U_{j-1}^i, \tag{23.7}$$

where $\nu := k/h^2$ is known as the *mesh ratio*.

Suppose that all approximate solution values at time level i, $\{U_j^i\}_{j=0}^{N_x}$, are known. Now note that $U_0^{i+1} = a((i+1)k)$ and $U_{N_x}^{i+1} = b((i+1)k)$ are given by the boundary conditions (23.4). Equation (23.7) then gives a formula for computing all other approximate values at time level $i + 1$, that is, $\{U_j^{i+1}\}_{j=1}^{N_x-1}$. Since we

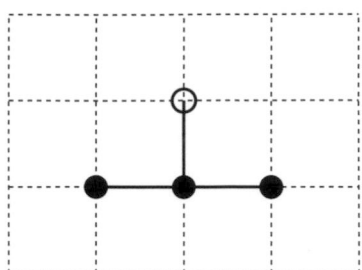

Fig. 23.3. Stencil for FTCS. Solid circles indicate the location of values that must be known in order to obtain the value located at the open circle.

are supplied with the time-zero values, $U_j^0 = g(jh)$ from (23.3), this means that the complete set of approximations $\{U_j^i\}_{j=0,i=0}^{N_x,\ N_t}$ can be computed by stepping forward in time. The method defined by (23.7) is known as FTCS, which stands for forward difference in **t**ime, **c**entral difference in **s**pace. Figure 23.3 illustrates the *stencil* for FTCS. Here, the solid circles indicate the location of values U_{j-1}^i, U_j^i and U_{j+1}^i that must be known in order to obtain the value U_j^{i+1} located at the open circle.

We may collect all the interior values at time level i into a vector,

$$\mathbf{U}^i := \begin{bmatrix} U_1^i \\ U_2^i \\ \vdots \\ \vdots \\ U_{N_x-1}^i \end{bmatrix} \in \mathbb{R}^{N_x-1}. \tag{23.8}$$

Exercise 23.3 then asks you to confirm that FTCS may be written

$$\mathbf{U}^{i+1} = F\mathbf{U}^i + \mathbf{p}^i, \qquad \text{for } 0 \leq i \leq N_t - 1, \tag{23.9}$$

with

$$\mathbf{U}^0 = \begin{bmatrix} g(h) \\ g(2h) \\ \vdots \\ \vdots \\ g((N_x-1)h) \end{bmatrix} \in \mathbb{R}^{N_x-1},$$

where the matrix F has the form

$$F = \begin{bmatrix} 1-2\nu & \nu & 0 & \cdots & \cdots & 0 \\ \nu & 1-2\nu & \nu & 0 & \ddots & \vdots \\ 0 & \ddots & \ddots & \ddots & \ddots & \vdots \\ \vdots & \ddots & \ddots & \ddots & \ddots & 0 \\ \vdots & & \ddots & \ddots & 1-2\nu & \nu \\ 0 & \cdots & \cdots & 0 & \nu & 1-2\nu \end{bmatrix} \in \mathbb{R}^{(N_x-1)\times(N_x-1)},$$

and the vector p^i has the form

$$p^i = \begin{bmatrix} \nu a(ik) \\ 0 \\ \vdots \\ \vdots \\ 0 \\ \nu b(ik) \end{bmatrix} \in \mathbb{R}^{N_x-1}.$$

Here, $F\mathbf{U}^i$ denotes a matrix–vector product.

Computational example Figure 23.4 illustrates a numerical solution produced by FTCS on the problem of Figure 23.1, with $T = 3$. We chose $N_x = 14$ and $N_t = 199$, so $h = \pi/14 \approx 0.22$ and $k = 3/199 \approx 0.015$, giving $\nu \approx 0.3$. The numerical solution appears to match the exact solution, shown in Figure 23.1. Computing the worst-case grid error, $\max_{0\le j\le N_x, 0\le i\le N_t} |U^i_j - u(jh, ik)|$, produced 0.0012, which confirms the close agreement. As can be seen from Figure 23.4, we used a grid where k is much smaller than h – we divided the x-axis into only 15 points, compared with 200 points on the t-axis. In Figure 23.5 we show what happens if we try to correct this imbalance. Here, we reduced N_t to 94, so $k \approx 0.032$ and $\nu \approx 0.63$. We see that the numerical solution has developed oscillations that render it useless as an approximation to $u(x, t)$. Taking smaller values of N_t, that is, larger timesteps k, leads to more dramatic oscillations. In Section 23.7 we develop some theory that explains this behaviour. We finish this section by deriving an alternative method that is more computationally expensive, but does not suffer from the type of instability seen in Figure 23.5. ◇

Replacing the forward difference in time in FTCS by a backward difference gives

$$k^{-1}\nabla_t U^i_j - h^{-2}\delta^2_x U^i_j = 0,$$

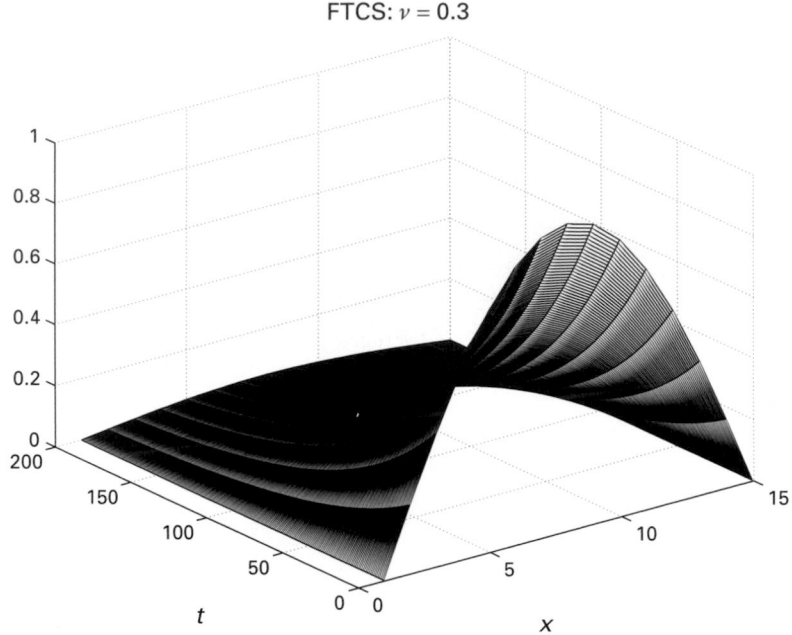

Fig. 23.4. FTCS solution on the heat equation (23.2), (23.3) and (23.4) with initial and boundary conditions (23.5). Here $N_x = 14$ and $N_t = 199$, so $v \approx 0.3$.

or, in more detail,

$$\frac{U_j^i - U_j^{i-1}}{k} - \frac{U_{j+1}^i - 2U_j^i + U_{j-1}^i}{h^2} = 0.$$

It is convenient to write this as a process that goes from time level i to $i + 1$, that is, to increase the time index by 1, which allows the method to be written

$$U_j^{i+1} = U_j^i + v\left(U_{j+1}^{i+1} - 2U_j^{i+1} + U_{j-1}^{i+1}\right). \tag{23.10}$$

The method defined by (23.10) is known as BTCS, which stands for **b**ackward difference in **t**ime, **c**entral difference in **s**pace. Figure 23.6 illustrates the stencil for BTCS. Unlike FTCS, with BTCS there is no explicit way to compute $\{U_j^{i+1}\}_{j=1}^{N_x-1}$ from $\{U_j^i\}_{j=1}^{N_x-1}$. Using the vector notation (23.8), Exercise 23.4 asks you to show that the recurrence (23.10) for BTCS may be written

$$B\mathbf{U}^{i+1} = \mathbf{U}^i + \mathbf{q}^i, \quad \text{for } 0 \le i \le N_t - 1, \tag{23.11}$$

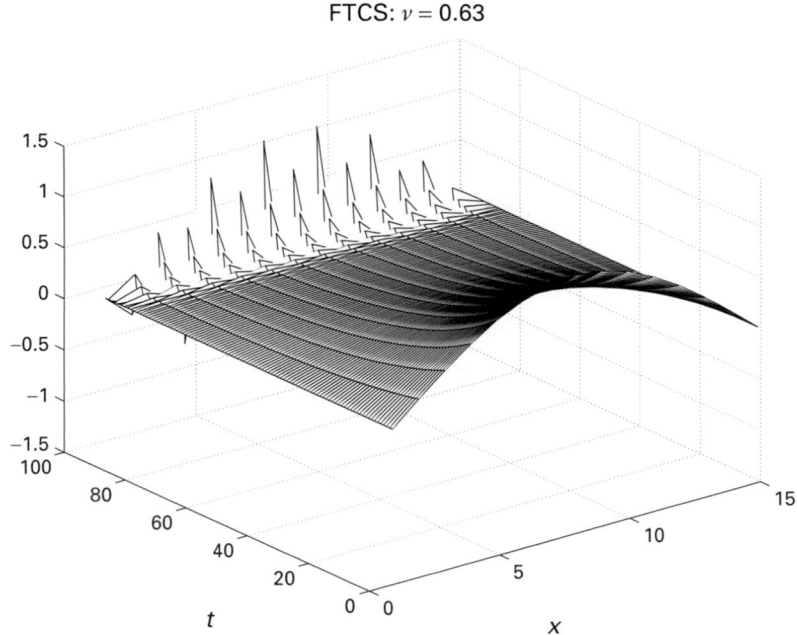

Fig. 23.5. FTCS solution on the heat equation (23.2), (23.3) and (23.4) with initial and boundary conditions (23.5). Here $N_x = 14$ and $N_t = 94$, so $v \approx 0.63$.

where the matrix B has the form

$$
B = \begin{bmatrix}
1+2v & -v & 0 & \cdots & & \cdots & 0 \\
-v & 1+2v & -v & 0 & & \ddots & \vdots \\
0 & \ddots & \ddots & \ddots & \ddots & & \vdots \\
\vdots & \ddots & \ddots & \ddots & \ddots & & 0 \\
\vdots & & \ddots & \ddots & -v & 1+2v & -v \\
0 & & \cdots & \cdots & 0 & -v & 1+2v
\end{bmatrix} \in \mathbb{R}^{(N_x-1)\times(N_x-1)},
$$

$$(23.12)$$

and the vector $\mathbf{q}^i$ has the form

$$
\mathbf{q}^i = \begin{bmatrix}
va((i+1)k) \\
0 \\
\vdots \\
\vdots \\
0 \\
vb((i+1)k)
\end{bmatrix} \in \mathbb{R}^{N_x-1}.
$$

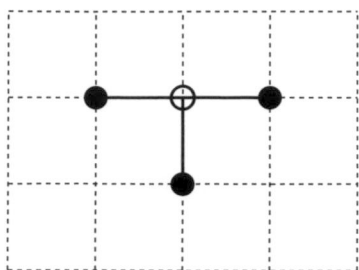

Fig. 23.6.   Stencil for BTCS. Solid circles indicate the location of values that must be known in order to obtain the value located at the open circle.

The formulation (23.11) reveals that, given U^i, we may compute U^{i+1} by solving a system of linear equations. This is a standard problem in numerical analysis, see Section 23.9 for references.

Computational example Figure 23.7 gives the BTCS numerical solution for the problem in Figure 23.1, with $T = 3$. We used $N_x = 14$ and $N_t = 9$, so $h = \pi/14 \approx 0.22$ and $k = 3/9 \approx 0.33$, giving $\nu \approx 6.6$. The numerical solution agrees qualitatively with the exact solution in Figure 23.1, and we found that the worst-case grid error, $\max_{0 \leq j \leq N_x, 0 \leq i \leq N_t} |U_j^i - u(jh, ik)|$, was a respectable 0.055.   ◇

23.6 Local accuracy

It is intuitively reasonable to judge the accuracy of a finite difference method by looking at the residual when the exact solution is substituted into the difference formula. For FTCS, letting u_j^i denote the exact solution $u(jh, ik)$, the *local accuracy* is defined to be

$$R_j^i := k^{-1}\Delta_t u_j^i - h^{-2}\delta_x^2 u_j^i.$$ (23.13)

Using the Taylor series results in Table 23.1, this may be expanded as

$$R_j^i = \left(\frac{\partial u}{\partial t} + \tfrac{1}{2}k\frac{\partial^2 u}{\partial t^2} + O(k^2)\right) - \left(\frac{\partial^2 u}{\partial x^2} + \tfrac{1}{12}h^2\frac{\partial^4 u}{\partial x^4} + O(h^4)\right),$$

where all functions $\partial u/\partial t$, $\partial^2 u/\partial t^2$, etc., are evaluated at $x = jh, t = ik$. Since u satisfies the PDE (23.2), we have

$$R_j^i = \tfrac{1}{2}k\frac{\partial^2 u}{\partial t^2} - \tfrac{1}{12}h^2\frac{\partial^4 u}{\partial x^4} + O(k^2) + O(h^4).$$ (23.14)

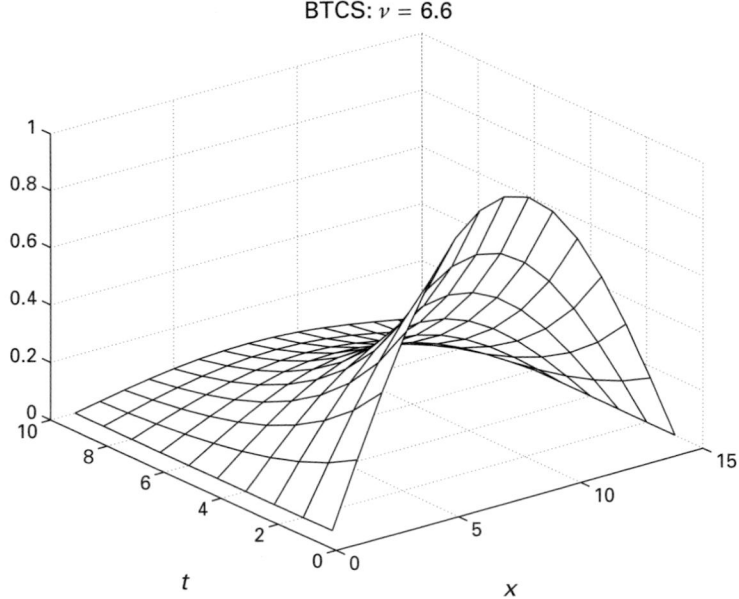

Fig. 23.7. BTCS solution on the heat equation (23.2), (23.3) and (23.4) with initial and boundary conditions (23.5). Here $N_x = 14$ and $N_t = 9$, so $\nu \approx 6.6$.

The expansion (23.14) shows that the local accuracy of FTCS behaves as $O(k) + O(h^2)$. Hence, FTCS may be described as *first order in time and second order in space*.

For BTCS, the local accuracy is defined as

$$R^i_j := k^{-1} \nabla_t u^i_j - h^{-2} \delta^2_x u^i_j. \tag{23.15}$$

In this case it is convenient to use Taylor series results from Table expansion about time level $(i + 1)k$, and we find that

$$R^i_j = -\tfrac{1}{2} k \frac{\partial^2 u}{\partial t^2} - \tfrac{1}{12} h^2 \frac{\partial^4 u}{\partial x^4} + O(k^2) + O(h^4), \tag{23.16}$$

with the functions evaluated at $x = jh$, $t = ik$. Exercise 23.5 asks you to fill in the details. This shows that BTCS has the same order of local accuracy as FTCS.

23.7 Von Neumann stability and convergence

A fundamental, and seemingly modest, requirement of a finite difference method is that of convergence – the error should tend to zero as k and h are decreased to zero. It turns out that convergence is quite a subtle issue. One aspect that must be addressed is the choice of norm in which convergence is measured; in the limit $k \to 0$, $h \to 0$, we are dealing with infinite-dimensional vector spaces, so we lose the property that 'all norms are equivalent'.

There is, however, a wonderful and very general result, known as the Lax Equivalence Theorem, which states that a method converges if and only if its local accuracy tends to zero as $k \to 0$, $h \to 0$ **and** it satisfies a stability condition. The particular stability condition to be satisfied depends on the norm in which convergence is measured. We do not have the space to go into any detail on this matter, but readers with a feel for Fourier analysis may appreciate that the following stability definition is related to the L_2 norm.

> **Definition** A finite difference method generating approximations U_j^i is *stable in the sense of von Neumann* if, ignoring initial and boundary conditions, under the substitution $U_j^i = \xi^i e^{\mathbf{i}\beta jh}$ it follows that[1] $|\xi| \le 1$ for all $\beta h \in [-\pi, \pi]$. Here $\mathbf{i}$ denotes the unit imaginary number. ◇

To illustrate the idea, taking FTCS in the form (23.7) and substituting $U_j^i = \xi^i e^{\mathbf{i}\beta jh}$ gives

$$\xi^{i+1} e^{\mathbf{i}\beta jh} = \nu \xi^i e^{\mathbf{i}\beta jh} e^{\mathbf{i}\beta h} + (1 - 2\nu)\xi^i e^{\mathbf{i}\beta jh} + \nu \xi^i e^{\mathbf{i}\beta jh} e^{-\mathbf{i}\beta h}.$$

So

$$\xi = \nu e^{\mathbf{i}\beta h} + (1 - 2\nu) + \nu e^{-\mathbf{i}\beta h}$$
$$= 1 + \nu \left(e^{\mathbf{i}\beta h} - 2 + e^{-\mathbf{i}\beta h} \right)$$
$$= 1 + \nu \left(e^{\mathbf{i}\frac{1}{2}\beta h} - e^{-\mathbf{i}\frac{1}{2}\beta h} \right)^2$$
$$= 1 + \nu \left(2\mathbf{i} \sin(\tfrac{1}{2}\beta h) \right)^2$$
$$= 1 - 4\nu \sin^2(\tfrac{1}{2}\beta h).$$

The condition $|\xi| \le 1$ thus becomes

$$|1 - 4\nu \sin^2(\tfrac{1}{2}\beta h)| \le 1,$$

which simplifies to

$$0 \le \nu \sin^2(\tfrac{1}{2}\beta h) \le \tfrac{1}{2}.$$

For $\beta h \in [-\pi, \pi]$ the quantity $\sin^2(\tfrac{1}{2}\beta h)$ takes values between 0 and 1, and hence stability in the sense of von Neumann for FTCS is equivalent to

$$\nu \le \tfrac{1}{2}. \tag{23.17}$$

---

[1] A more general definition allows $|\xi| \le 1 + Ck$ for some constant C, but our simpler version suffices here.

Returning to our previous computations, we see that a stable value of $v \approx 0.3$ was used for FTCS in Figure 23.4, whereas Figure 23.5 went beyond the stability limit, with $v \approx 0.63$. In practice, FCTS is only useful for $v \leq \frac{1}{2}$. If we consider refining the grid, that is reducing h and k to get more accuracy, then we do so while respecting this condition. It is typical to choose v, say $v = 0.45$, and consider the limit $h \to 0$ with fixed mesh ratio $k/h^2 = v$. In this regime, k tends to zero much more quickly than h.

Exercise 23.6 asks you to show that BTCS is *unconditionally stable*, that is, stability in the sense of von Neumann is guaranteed for all $v > 0$. This is consistent with Figure 23.7, where a relatively large value of v did not give rise to any instabilities.

23.8 Crank–Nicolson

We have seen that FTCS and BTCS are both of local accuracy $O(k) + O(h^2)$. The $O(k)$ accuracy in time arises from the use of first order forward or backward differencing in time. The Crank–Nicolson method uses a clever trick to achieve second order in time without the need to deal with more than two time levels.

To derive the Crank–Nicolson method, we temporarily entertain the idea of an intermediate time level at $(i + \frac{1}{2})k$. The heat equation (23.2) may then be approximated by

$$k^{-1}\delta_t U_j^{i+\frac{1}{2}} - h^{-2}\delta_x^2 U_j^{i+\frac{1}{2}} = 0.$$

This finite difference formula has an appealing symmetry. However, we have introduced points that are not on the grid. We may overcome this difficulty by applying the time averaging operator, μ_t, on the right-hand term, to get a new method

$$k^{-1}\delta_t U_j^{i+\frac{1}{2}} - h^{-2}\delta_x^2 \mu_t U_j^{i+\frac{1}{2}} = 0,$$

that is

$$k^{-1}(U_j^{i+1} - U_j^i) - h^{-2}\delta_x^2 \tfrac{1}{2}(U_j^{i+1} + U_j^i) = 0.$$

This may be written as

$$2(1 + v)U_j^{i+1} = vU_{j+1}^{i+1} + vU_{j-1}^{i+1} + vU_{j+1}^i + 2(1 - v)U_j^i + vU_{j-1}^i. \quad (23.18)$$

This is Crank–Nicolson. The stencil is shown in Figure 23.8. Because of its inherent symmetry, the method has local accuracy $O(k^2) + O(h^2)$. Exercise 23.8 asks you to confirm this. Crank–Nicolson has two features in common with BTCS. First, it is implicit, requiring a system of linear equations to be solved in order to compute $\mathbf{U}^{i+1}$ from $\mathbf{U}^i$. The equations may be written

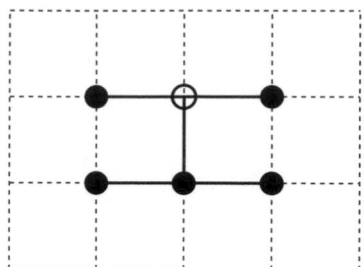

Fig. 23.8.   Stencil for Crank–Nicolson. Solid circles indicate the location of values that must be known in order to obtain the value located at the open circle.

$$\widehat{B}\mathbf{U}^{i+1} = \widehat{F}\mathbf{U}^i + \mathbf{r}^i, \quad \text{for } 0 \leq i \leq N_t - 1, \tag{23.19}$$

where the matrices $\widehat{B}$ and $\widehat{F}$ have the form

$$\widehat{B} = \begin{bmatrix} 1+v & -\frac{1}{2}v & 0 & \cdots & \cdots & 0 \\ -\frac{1}{2}v & 1+v & -\frac{1}{2}v & 0 & \ddots & \vdots \\ 0 & \ddots & \ddots & \ddots & \ddots & \vdots \\ \vdots & \ddots & \ddots & \ddots & \ddots & 0 \\ \vdots & \ddots & \ddots & -\frac{1}{2}v & 1+v & -\frac{1}{2}v \\ 0 & \cdots & \cdots & 0 & -\frac{1}{2}v & 1+v \end{bmatrix} \in \mathbb{R}^{(N_x-1)\times(N_x-1)},$$

$$\widehat{F} = \begin{bmatrix} 1-v & \frac{1}{2}v & 0 & \cdots & \cdots & 0 \\ \frac{1}{2}v & 1-v & \frac{1}{2}v & 0 & \ddots & \vdots \\ 0 & \ddots & \ddots & \ddots & \ddots & \vdots \\ \vdots & \ddots & \ddots & \ddots & \ddots & 0 \\ \vdots & \ddots & \ddots & \frac{1}{2}v & 1-v & \frac{1}{2}v \\ 0 & \cdots & \cdots & 0 & \frac{1}{2}v & 1-v \end{bmatrix} \in \mathbb{R}^{(N_x-1)\times(N_x-1)},$$

and the vector $\mathbf{r}^i$ has the form

$$\mathbf{r}^i = \begin{bmatrix} \frac{1}{2}v\,(a(ik) + a((i+1)k)) \\ 0 \\ \vdots \\ \vdots \\ 0 \\ \frac{1}{2}v\,(b(ik) + b((i+1)k)) \end{bmatrix} \in \mathbb{R}^{N_x-1},$$

see Exercise 23.9. Second, it is stable in the sense of von Neumann for all $\nu > 0$, see Exercise 23.10. The extra order of local accuracy in time makes it a popular choice. Exercise 23.11 gives an alternative derivation of the method.

Computational example Recall that the BTCS computation in Figure 23.7 produced a worst-case grid error of 0.055. Switching to Crank–Nicolson, we find that the error reduces to 0.0019, which reflects the higher order of local accuracy in time. ◇

23.9 Notes and references

This chapter was designed to give only the most cursory introduction to finite differences. Excellent, accessible texts that give much more detail and, in particular, describe methods for solving the linear systems such as (23.11) and (23.19), and also do justice to the Lax Equivalence Theorem, include (Iserles, 1996; Mitchell and Griffiths, 1980; Morton and Mayers, 1994; Strikwerda, 1989). A freely available work of similarly high quality is the unpublished text, *Finite Difference and Spectral Methods for Ordinary and Partial Differential Equations*, 1996, by Lloyd N. Trefethen, which is downloadable from http://web.comlab.ox.ac.uk/oucl/work/nick.trefethen/pdetext.html.

Details of how to transform the Black–Scholes PDE (8.15) into standard heat equation form (23.2) can be found, for example, in (Nielsen, 1999, Section 6.7) and (Wilmott *et al.*, 1995, Section 5.4).

Finite difference methods represent the most conceptually straightforward approach to solving a PDE numerically, and they appear to be the most popular choice in the mathematical finance community. However, it is worth pointing out that there are other areas of science and engineering where numerical methods for PDEs have reached a greater level of maturity, and in many cases other techniques, most notably *finite element* methods, have found considerable favour.

EXERCISES

23.1. ⋆⋆ Show that $\Delta \nabla = \nabla \Delta$; that is, for any sequence $\{y_m\}$, $\Delta \nabla y_m = \nabla \Delta y_m$. Similarly, establish the following identities relating finite difference operators:

$$\Delta \nabla = \nabla - \Delta,$$
$$\Delta \nabla = \delta^2,$$
$$\Delta_0 = \mu \delta,$$
$$\Delta_0 = \delta \mu,$$

$$\Delta^2 = \delta^2 E,$$
$$\Delta^2 = E\delta^2.$$

23.2. ⋆⋆ Verify the Taylor series expansions in Table 23.1.

23.3. ⋆ Verify that FTCS, (23.7), may be written in the form (23.9).

23.4. ⋆ Verify that BTCS, (23.10), may be written in the form (23.11).

23.5. ⋆⋆ Using Table 23.1, show that the local accuracy of BTCS, defined in (23.15), satisfies (23.16).

23.6. ⋆ ⋆ ⋆ By copying the analysis that led to (23.17), show that BTCS is stable in the sense of von Neumann for all $\nu > 0$.

23.7. ⋆⋆ Show that Crank–Nicolson, (23.18), can be expressed as

$$\left(1 - \tfrac{1}{2}\nu\delta_x^2\right) U_j^{i+1} = \left(1 + \tfrac{1}{2}\nu\delta_x^2\right) U_j^i.$$

23.8. ⋆ ⋆ ⋆ By analogy with (23.13) and (23.15), define the local accuracy for Crank–Nicolson and show that it is $O(k^2) + O(h^2)$.

23.9. ⋆⋆ Verify that Crank–Nicolson, (23.18), may be written in the form (23.19).

23.10. ⋆ ⋆ ⋆ Show that a von Neumann stability analysis of Crank–Nicolson, (23.18) leads to

$$\xi = \frac{1 - 2\nu \sin^2(\tfrac{1}{2}\beta h)}{1 + 2\nu \sin^2(\tfrac{1}{2}\beta h)}.$$

Deduce that the method is stable for all $\nu > 0$.

23.11. ⋆⋆ Suppose we take the average of the FTCS equation (23.9) and the BTCS equation (23.11) to get

$$\tfrac{1}{2}(I + B)\mathbf{U}^{i+1} = \tfrac{1}{2}(I + F)\mathbf{U}^i + \tfrac{1}{2}(\mathbf{p}^i + \mathbf{q}^i).$$

Show that this method is Crank–Nicolson. (The second order accuracy in time may now be understood by observing that averaging the local accuracy expansions (23.14) and (23.16) causes the $O(k)$ term to vanish.)

23.10 Program of Chapter 23 and walkthrough

The program ch23 implements BTCS for the heat equation (23.2) with initial and boundary conditions (23.5), and plots the solution in the style of Figure 23.7. It is listed in Figure 23.9. After initializing parameters, we set up the Nx-1 by Nx-1 array B, which has the form displayed in (23.12).

```
%CH23    Program for Chapter 23
%
% Backward time central space (BTCS) for heat eqn

clf
%%%%%%%%%%%%% Parameters %%%%%%%%%%%%%%%%%
L = pi; Nx = 9; dx = L/Nx;
T = 3; Nt = 19; dt = T/Nt; nu = dt/dx^2;
%%%%%%%%%%%%%%%%%%%%%%%%%%%%%%%%%%%%%%%%%%%%%

B = (1+2*nu)*eye(Nx-1,Nx-1) - nu*diag(ones(Nx-2,1),1) - nu*diag(ones(Nx-2,1),-1);
U = zeros(Nx-1,Nt+1);
U(:,1) = sin([dx:dx:L-dx]');
for i = 1:Nt
   x = B\U(:,i);
   U(:,i+1) = x;
end
bc = zeros(1,Nt+1);
U = [bc;U;bc];

mesh(U')
xlabel('x','FontSize',20')
ylabel('t','FontSize',20')
```

Fig. 23.9.   Program of Chapter 23: ch23.m.

This is done with eye, diag and ones. The command eye(Nx-1,Nx-1) sets up an identity matrix

$$
\begin{bmatrix}
1 & 0 & \cdots & \cdots & \cdots & 0 \\
0 & 1 & 0 & \cdots & \cdots & 0 \\
\vdots & 0 & \ddots & \ddots & \ddots & \vdots \\
\vdots & \ddots & \ddots & \ddots & \ddots & \vdots \\
\vdots & \ddots & \ddots & \ddots & \ddots & 0 \\
0 & \cdots & \cdots & \cdots & 0 & 1
\end{bmatrix}
\in \mathbb{R}^{(N_x-1)\times(N_x-1)}.
$$

The array

$$
\begin{bmatrix}
1 \\
1 \\
\vdots \\
\vdots \\
1
\end{bmatrix}
\in \mathbb{R}^{N_x-2}
$$

is created by ones(Nx-2,1) and used in the diag function. Generally, diag(v,k) creates a two-dimensional array with v placed down the kth sub-/super-diagonal and zeros elsewhere. In our case, `diag(ones(Nx-2, 1),1)` and `diag(ones(Nx-2, 1),-1)` correspond to

$$
\begin{bmatrix}
0 & 1 & 0 & \cdots & \cdots & 0 \\
0 & 0 & 1 & 0 & \cdots & 0 \\
0 & \ddots & \ddots & \ddots & \ddots & \vdots \\
\vdots & \ddots & \ddots & \ddots & \ddots & 0 \\
\vdots & \ddots & \ddots & \ddots & \ddots & 1 \\
0 & \cdots & \cdots & 0 & 0 & 0
\end{bmatrix}
\quad \text{and} \quad
\begin{bmatrix}
0 & 0 & \cdots & \cdots & \cdots & 0 \\
1 & 0 & \ddots & \ddots & \cdots & 0 \\
0 & 1 & 0 & \ddots & \ddots & \vdots \\
\vdots & \ddots & \ddots & \ddots & \ddots & \vdots \\
\vdots & \ddots & \ddots & \ddots & \ddots & 0 \\
0 & \cdots & \cdots & 0 & 1 & 0
\end{bmatrix}
\in \mathbb{R}^{(N_x-1)\times(N_x-1)}.
$$

respectively. The `Nx-1` by `Nx-1` array `U` is used to store the numerical solution; successive columns hold the solution $\mathbf{U}^i$ in (23.8) at successive time levels. The initial condition is inserted into the first column with `U(:,1) = sin([dx:dx:L-dx]');`. We then enter a `for` loop that steps forward in time. Generally, if `A` and `b` are compatible two- and one-dimensional arrays, respectively, then `A\b` computes the solution `x` to the linear system `A*x = b`. It follows that the line `x = B\U(:,i);` solves the required system (23.11), and `U(:,i+1) = x;` assigns this solution to the next column of `U`. Note that $\mathbf{q}^i \equiv 0$ in (23.11) because of the zero boundary conditions. The line `U = [bc;U;bc];` pads out `U` by adding a row of zeros at the top and bottom, corresponding to those zero boundary conditions.

PROGRAMMING EXERCISES

P23.1. Using colon subarray notation, as in `ch16`, or otherwise, alter `ch23` so that FTCS is used. Toy with the stability constraint $\nu \leq \frac{1}{2}$.

P23.2. Implement Crank–Nicolson on the heat equation and compare its accuracy with that of FTCS and BTCS.

Quotes

In order to solve this differential equation
you look at it till a solution occurs to you.
    GEORGE POLYÁ, 1887–1985, source http://math.furman.edu/-mwoodard/mquot.html

Numerical theory for PDEs of evolution is sometimes presented in a deceptively simple way.
On the face of it, nothing could be more straightforward:
discretize all spatial derivatives by finite differences
and apply a reputable ODE solver,
without paying heed to the fact that, actually,
one is attempting to solve a PDE.
This nonsense has, unfortunately, taken root in many textbooks and lecture courses,
which, not to mince words, propagate shoddy mathematics and poor numerical practice.
Reputable literature is surprisingly scarce,
considering the importance and depth of the subject.
    ARIEH ISERLES (Iserles, 1996)

Spelling note #1: the name is 'Nicolson', not 'Nicholson'.

LLOYD N. TREFETHEN, *Finite Difference and Spectral Methods for Ordinary and Partial Differential Equations*, 1996; see Section 23.9.

24

Finite difference methods for the Black–Scholes PDE

24.1 Motivation

The previous chapter introduced finite difference methods. Here, we apply this idea to the Black–Scholes PDE. This is not entirely straightforward because the PDE is slightly more general than the heat equation used in Chapter 23 and the boundary conditions are not quite so convenient.

24.2 FTCS, BTCS and Crank–Nicolson for Black–Scholes

The Black–Scholes PDE (8.15) is typically augmented with a *final time condition* – examples that we have seen include (8.16), (8.25), (17.1) and (19.2). Since convention (and every book on numerical PDEs) dictates that problems should be specified in *initial time condition* form, we make the change of variable $\tau = T - t$. In this way τ represents the time to expiry and runs from T to 0 when t runs from 0 to T. Under this transformation the Black–Scholes PDE (8.15) becomes

$$\frac{\partial V}{\partial \tau} - \tfrac{1}{2}\sigma^2 S^2 \frac{\partial^2 V}{\partial S^2} - rS\frac{\partial V}{\partial S} + rV = 0. \tag{24.1}$$

In this section we focus on European calls and puts. The $t = T$ condition for a European call, (8.16), becomes the $\tau = 0$ condition

$$C(S, 0) = \max(S(0) - E, 0). \tag{24.2}$$

Similarly, the European put condition (8.25) changes to

$$P(S, 0) = \max(E - S(0), 0). \tag{24.3}$$

Turning to boundary conditions, the European call and put involve the PDE on the domain $S \in [0, \infty]$. This presents a difficulty. We must represent this range by a finite set of points. A reasonable fix is to truncate the domain to $S \in [0, L]$, where L is some suitably large value. Using (8.17) and (8.18), this gives call boundary conditions

$$C(0, \tau) = 0 \quad \text{and} \quad C(L, \tau) = L. \tag{24.4}$$

Similarly, from (8.26) and (8.27) we obtain

$$P(0, \tau) = Ee^{-r\tau} \quad \text{and} \quad P(L, \tau) = 0 \tag{24.5}$$

for a European put.

We are now able to use a grid $\{jh, ik\}_{j=0, i=0}^{N_x, N_t}$, as shown in Figure 23.2. Letting

$$\mathbf{V}^i := \begin{bmatrix} V_1^i \\ V_2^i \\ \vdots \\ \vdots \\ V_{N_x-1}^i \end{bmatrix} \in \mathbb{R}^{N_x-1}$$

denote the numerical solution at time level i, we have $\mathbf{V}^0$ specified by the initial data (24.2) or (24.3) and the boundary values V_0^i and $V_{N_x}^i$ for all $1 \le i \le N_t$ specified by the boundary conditions (24.4) or (24.5).

To obtain a generalized version of FTCS for the PDE (24.1) we use the full central difference operator from Table 23.1 for the $\partial V / \partial S$ term and evaluate the V term at (jh, ik) to get the difference equation

$$\frac{V_j^{i+1} - V_j^i}{k} - \tfrac{1}{2}\sigma^2 (jh)^2 \frac{\left(V_{j+1}^i - 2V_j^i + V_{j-1}^i\right)}{h^2} - rjh \left(\frac{V_{j+1}^i - V_{j-1}^i}{2h}\right) + rV_j^i = 0. \tag{24.6}$$

The corresponding generalization of BTCS is

$$\frac{V_j^{i+1} - V_j^i}{k} - \tfrac{1}{2}\sigma^2 (jh)^2 \frac{\left(V_{j+1}^{i+1} - 2V_j^{i+1} + V_{j-1}^{i+1}\right)}{h^2}$$
$$- rjh \left(\frac{V_{j+1}^{i+1} - V_{j-1}^{i+1}}{2h}\right) + rV_j^{i+1} = 0. \tag{24.7}$$

The matrix–vector representation of FTCS in (23.9) remains valid if we re-define

$$F = (1 - rk)I + \tfrac{1}{2}k\sigma^2 D_2 T_2 + \tfrac{1}{2}kr D_1 T_1$$

and

$$
\mathbf{p}^i =
\begin{bmatrix}
\frac{1}{2}k(\sigma^2 - r)V_0^i \\
0 \\
\vdots \\
\vdots \\
0 \\
\frac{1}{2}k(N_x - 1)(\sigma^2(N_x - 1) + r)V_{N_x}^i
\end{bmatrix},
$$

where

$$
D_1 =
\begin{bmatrix}
1 & 0 & \cdots & \cdots & 0 \\
0 & 2 & 0 & \ddots & \vdots \\
\vdots & 0 & 3 & \ddots & \vdots \\
\vdots & \ddots & \ddots & \ddots & 0 \\
0 & \cdots & \cdots & 0 & N_x - 1
\end{bmatrix},
\quad
D_2 =
\begin{bmatrix}
1^2 & 0 & \cdots & \cdots & 0 \\
0 & 2^2 & 0 & \ddots & \vdots \\
\vdots & 0 & 3^2 & \ddots & \vdots \\
\vdots & \ddots & \ddots & \ddots & 0 \\
0 & \cdots & \cdots & 0 & (N_x - 1)^2
\end{bmatrix}
$$

and

$$
T_1 =
\begin{bmatrix}
0 & 1 & 0 & \cdots & \cdots & 0 \\
-1 & 0 & 1 & \ddots & \ddots & \vdots \\
0 & \ddots & \ddots & \ddots & \ddots & \vdots \\
\vdots & \ddots & \ddots & \ddots & \ddots & 0 \\
\vdots & \ddots & \ddots & -1 & 0 & 1 \\
0 & \cdots & \cdots & 0 & -1 & 0
\end{bmatrix},
\quad
T_2 =
\begin{bmatrix}
-2 & 1 & 0 & \cdots & \cdots & 0 \\
1 & -2 & 1 & \ddots & \ddots & \vdots \\
0 & 1 & \ddots & \ddots & \ddots & \vdots \\
\vdots & \ddots & \ddots & \ddots & \ddots & 0 \\
\vdots & \ddots & \ddots & 1 & -2 & 1 \\
0 & \cdots & \cdots & 0 & 1 & -2
\end{bmatrix}.
$$

Similarly, BTCS has the form (23.11) with

$$
B = (1 + rk)I - \tfrac{1}{2}k\sigma^2 D_2 T_2 - \tfrac{1}{2}kr D_1 T_1
$$

and

$$
\mathbf{q}^i =
\begin{bmatrix}
\frac{1}{2}k(\sigma^2 - r)V_0^{i+1} \\
0 \\
\vdots \\
\vdots \\
0 \\
\frac{1}{2}k(N_x - 1)(\sigma^2(N_x - 1) + r)V_{N_x}^{i+1}
\end{bmatrix},
$$

see Exercise 24.1.

One way to generalize the Crank–Nicolson scheme (23.18) is to adopt the viewpoint of Exercise 23.11 and take the average of the FTCS and BTCS formulas

(23.9) and (23.11) to give

$$\tfrac{1}{2}(I + B)\mathbf{V}^{i+1} = \tfrac{1}{2}(I + F)\mathbf{V}^i + \tfrac{1}{2}(\mathbf{p}^i + \mathbf{q}^i). \tag{24.8}$$

Computational example We used our three finite difference methods to value a European put option with parameters $E = 4$, $\sigma = 0.3$, $r = 0.03$ and $T = 1$. We truncated the asset range at $L = 10$. Since the exact value is known from the Black–Scholes formula (8.24), we may check the error. We focused on the maximum error at time zero:

$$\text{err}^0 := \max_{1 \le j \le N_x - 1} |V_j^{N_t} - V(jh, \tau = T)|. \tag{24.9}$$

With $N_x = 50$ and $N_t = 500$, so $k = 2 \times 10^{-3}$ and $h = 0.2$, we found that $\text{err}^0 = 1.5 \times 10^{-3}$ for FTCS and $\text{err}^0 = 1.7 \times 10^{-3}$ for BTCS. With Crank–Nicolson we were able to reduce N_t to 50, so $k = 2 \times 10^{-2}$, and still get a comparable error, $\text{err}^0 = 1.6 \times 10^{-3}$. $\diamond$

Our treatment of stability and convergence of finite difference methods in Chapter 23 does not carry through directly to this section, since the PDE (24.1) has nonconstant coefficients and includes a first order spatial derivative. However, similar conclusions may be drawn; see Section 24.5.

24.3 Down-and-out call example

To illustrate the flexibility of finite difference methods, we turn to the down-and-out call defined in Section 19.2. We know that the PDE holds for $B \le S$. Hence, we may truncate this to $B \le S \le L$ and use a grid of the form $\{B + jh, ik\}_{j=0, i=0}^{N_x, N_t}$, where $h = (L - B)/N_x$. The FTCS scheme (24.6) becomes

$$\frac{V_j^{i+1} - V_j^i}{k} - \tfrac{1}{2}\sigma^2 (B + jh)^2 \frac{\left(V_{j+1}^i - 2V_j^i + V_{j-1}^i\right)}{h^2}$$
$$- r(B + jh)\left(\frac{V_{j+1}^i - V_{j-1}^i}{2h}\right) + rV_j^i = 0$$

and the corresponding BTCS version is

$$\frac{V_j^{i+1} - V_j^i}{k} - \tfrac{1}{2}\sigma^2 (B + jh)^2 \frac{\left(V_{j+1}^{i+1} - 2V_j^{i+1} + V_{j-1}^{i+1}\right)}{h^2}$$
$$- r(B + jh)\left(\frac{V_{j+1}^{i+1} - V_{j-1}^{i+1}}{2h}\right) + rV_j^{i+1} = 0.$$

As before, these may be written in the matrix–vector forms (23.9) and (23.11), and the Crank–Nicolson method is given by (24.8).

The $\tau = 0$ condition (19.2) specifies $V_j^0 = \max(B + jh - E, 0)$ and the left-hand boundary condition (19.1) gives $V_0^i = 0$. At the right-hand boundary, a reasonable approach is to argue that, since S is large, the asset is very unlikely to hit the out barrier, so $V_{N_x}^i = C(L, \tau)$ may be imposed, where $C(S, t)$ denotes the European call value.

Computational example For the case $B = 2$, $E = 4$, $\sigma = 0.3$, $r = 0.03$ and $T = 1$ we used Crank–Nicolson to value a down-and-out call. In this case the exact solution (19.3) may be used to check the error. With the asset domain truncated at $L = 10$, and with $N_x = N_t = 50$, we found the maximum time-zero error (24.9) to be $\text{err}^0 = 1.1 \times 10^{-3}$. ◇

24.4 Binomial method as finite differences

Looking back to Chapter 16, we see some similarities between the binomial and finite difference methods:

- both work with discretizations of the time and asset domains,
- both advance in the time direction,
- both are designed to be more accurate as the discretization is refined.

The binomial method works in backward time – starting with option values at $t = T$ and finishing with a value at $t = 0$ and $S = S_0$. The finite difference methods are more general, in that they produce option values at **all** grid-points $\{jh, ik\}$; in particular, at time zero, option values are available for all initial asset prices $0, h, 2h, \ldots, L$. Nevertheless, it should seem plausible that the binomial method may be regarded as some explicit finite difference scheme that has been customized to produce a single time-zero option value. In this section we explain how the connection can be made concrete.

Starting with (8.15), we make the transformation $X = \log S$, which produces the constant coefficient PDE

$$\frac{\partial V}{\partial t} + \tfrac{1}{2}\sigma^2 \frac{\partial^2 V}{\partial X^2} + (r - \tfrac{1}{2}\sigma^2)\frac{\partial V}{\partial X} - rV = 0.$$

We then let $V = e^{rt}W$. This has the effect of eliminating the zeroth derivative term, to give

$$\frac{\partial W}{\partial t} + \tfrac{1}{2}\sigma^2 \frac{\partial^2 W}{\partial X^2} + (r - \tfrac{1}{2}\sigma^2)\frac{\partial W}{\partial X} = 0, \tag{24.10}$$

see Exercise 24.4.

Now, applying a backward difference formula for the time derivative and central differences for the space derivatives in (24.10) leads to the finite difference formula

$$\frac{W_j^{i+1} - W_j^i}{k} + \tfrac{1}{2}\sigma^2 \left(\frac{W_{j+1}^{i+1} - 2W_j^{i+1} + W_{j-1}^{i+1}}{h^2} \right)$$

$$+ (r - \tfrac{1}{2}\sigma^2) \left(\frac{W_{j+1}^{i+1} - W_{j-1}^{i+1}}{2h} \right) = 0. \qquad (24.11)$$

Setting $h^2 = \sigma^2 k$ has the effect of eliminating the W_j^{i+1} terms in (24.11), and the formula then reduces to

$$W_j^i = p^\star W_{j+1}^{i+1} + (1 - p^\star)W_{j-1}^{i+1}, \qquad (24.12)$$

where $p^\star = \tfrac{1}{2}\left(1 + \sqrt{k}(r/\sigma - \sigma/2)\right)$. Transforming back to V we find that

$$V_j^i = e^{-rk} \left(p^\star V_{j+1}^{i+1} + (1 - p^\star)V_{j-1}^{i+1} \right). \qquad (24.13)$$

Comparing (24.13) and (16.3), we see that the binomial method corresponds to using an explicit finite difference method on a transformed version of the Black–Scholes PDE. The finite differences are applied on a sub-grid, as illustrated in Figure 24.1. The coupling $h^2 = \sigma^2 k$ puts the method on the very cusp of von Neumann instability, see Exercise 24.5, which explains the undesirable but noncatastrophic oscillations observed in Section 16.4.

24.5 Notes and references

As we mentioned in Chapter 23, it is possible to convert the Black–Scholes PDE for European calls and puts into the heat equation form (23.2). Hence, it is perfectly reasonable to convert to that form before applying a finite difference method. We showed how to work directly with the Black–Scholes version (in reverse time) because in the case of more complicated options such a transformation may not be possible. We chose to discretize the spatial first derivative $\partial V/\partial S$ in (24.1) by a central difference. An alternative that is better in the case where the volatility is very small is *upwind* differencing; see (Iserles, 1996; Mitchell and Griffiths, 1980; Morton and Mayers, 1994; Strikwerda, 1989).

The texts (Clewlow and Strickland, 1998; Kwok, 1998; Wilmott, 1998; Wilmott *et al.*, 1995; Seydel, 2002) are good sources for more details about the application of finite differences to option valuation.

We saw in Chapter 18 that the problem of valuing an American option can be couched in terms of a linear complementarity problem. It is possible to develop

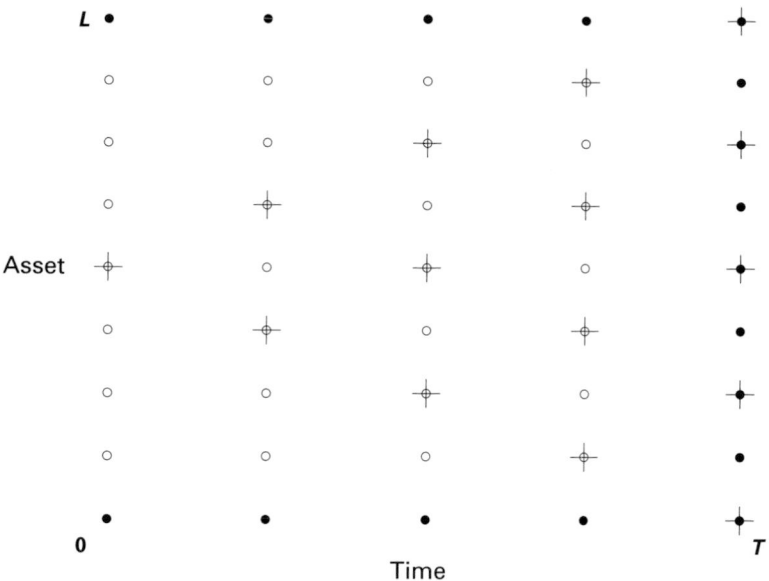

Fig. 24.1. An example of a finite difference grid $\{jh, ik\}_{j=0,i=0}^{N_x, N_t}$. Crosses mark points used by the binomial method (24.13) to obtain a single time-zero option value.

finite difference schemes for such problems; see (Wilmott *et al.*, 1995), for example. A promising, but often overlooked, alternative is to use a penalty method. Indeed, the basic binomial method of Chapter 18 is an example of a simple, explicit penalty method. More accurate versions are developed and analysed in (Forsyth and Vetzal, 2002). Our illustration in Section 24.4 of the connection between binomial and finite difference methods was based on Appendix C of (Forsyth and Vetzal, 2002). A fuller treatment of this topic can be found in (Kwok, 1998).

It is worth making the point that the development and implementation of numerical methods for PDEs is an area where a beginner is generally best advised to make use of existing technology: 'off the shelf' is preferable to 'roll your own'. However, a basic understanding of the nature of simple numerical methods, at the level of these last two chapters, gives a good feel for what to expect from PDE solvers.

MATLAB comes with a fairly simple built-in PDE solver, pdepe, and may be augmented with a PDE toolbox. Generally, there is an abundance of numerical PDE software available, both commercially and in the public domain. Good places to start are the Netlib Repository www.netlib.org/liblist.html and the Differential Equations and Related Topics page http://www.maths.dundee.ac.uk/software/index.html#DEs maintained by David Griffiths at the University of Dundee.

```
%CH24    Program for Chapter 24
%
%    Crank-Nicolson for a European put

clf

%%%%%% Problem and method parameters %%%%%%%
E = 4; sigma = 0.3; r = 0.03; T = 1;
L = 10; Nx = 50; Nt = 50; k = T/Nt; h = L/Nx;
%%%%%%%%%%%%%%%%%%%%%%%%%%%%%%%%%%%%%%%%

T1 =  diag(ones(Nx-2,1),1) - diag(ones(Nx-2,1),-1);
T2 = -2*eye(Nx-1,Nx-1)  + diag(ones(Nx-2,1),1) + diag(ones(Nx-2,1),-1);
mvec = [1:Nx-1];
D1 = diag(mvec);
D2 = diag(mvec.^2);
F = (1-r*k)*eye(Nx-1,Nx-1) + 0.5*k*sigma^2*D2*T2 + 0.5*k*r*D1*T1;
B = (1+r*k)*eye(Nx-1,Nx-1) - 0.5*k*sigma^2*D2*T2 - 0.5*k*r*D1*T1;
A1 = 0.5*(eye(Nx-1,Nx-1) + F);
A2 = 0.5*(eye(Nx-1,Nx-1) + B);

U = zeros(Nx-1,Nt+1);
U(:,1) = max(E-[h:h:L-h]',0);

for i = 1:Nt
   tau = (i-1)*k;
   p1 = k*(0.5*sigma^2 - 0.5*r)*E*exp(-r*(tau));
   q1 = k*(0.5*sigma^2 - 0.5*r)*E*exp(-r*(tau+k));
   rhs = A1*U(:,i) + [0.5*(p1+q1); zeros(Nx-2,1)];
   X = A2\rhs;
   U(:,i+1) = X;
end

bca = E*exp(-r*[0:k:T]);
bcb = zeros(1,Nt+1);
U = [bca;U;bcb];
mesh([0:k:T],[0:h:L],U)
xlabel('T-t'), ylabel('S'), zlabel('Put Value')
```

Fig. 24.2.   Program of Chapter 24: ch24.m.

EXERCISES

24.1. ⋆ Confirm that FTCS in (24.6) and BTCS in (24.7) have matrix–vector forms (23.9) and (23.11), respectively, as indicated in Section 24.2.

24.2. ⋆⋆ In the case of a European call option, point out a contradiction in the initial and boundary conditions (24.2) and (24.4). How could this be overcome?

24.3. ⋆⋆ Write the FTCS, BTCS and Crank–Nicolson methods for a down-and-out call option in matrix–vector form.

24.4. ⋆ ⋆ ⋆ Confirm that the transformations given in Section 24.4 convert (8.15) to (24.10).

24.5. ⋆⋆ Suppose that a constant diffusion coefficient, $\frac{1}{2}\sigma^2$, is introduced into the heat equation (23.2) to give

$$\frac{\partial u}{\partial t} = \frac{1}{2}\sigma^2 \frac{\partial^2 u}{\partial x^2}.$$

The FTCS method would then use

$$k^{-1}\Delta_t U_j^i - \frac{1}{2}h^{-2}\delta_x^2 U_j^i = 0.$$

Show that the von Neumann stability condition takes the form $\sigma^2 k \le h^2$.

24.6 Program of Chapter 24 and walkthrough

Our program ch24 implements Crank–Nicolson, (24.8), for a European put, producing a picture like that in Figure 11.4. It is listed in Figure 24.2. The structure of the code is similar to ch23, and the commands used have been explained in previous chapters.

PROGRAMMING EXERCISES

P24.1. Alter ch24 so that it values a down-and-out call option.

P24.2. Investigate the use of MATLAB's built-in PDE solver pdepe for option valuation. Type help pdepe or consult (Higham and Higham, 2000, Section 12.4) for details of how to use pdepe.

Quote

. . . one reason I've found financial engineering so exciting
is that banks pay attention to a lot of academic work.
In that sense, it's a very aggressive area,
because if you have a new method for solving a problem of interest,
there will be listeners.
And they'll come back, ask questions, be on the phone,
and fill the seminar room.

TOM COLEMAN, *Financial Engineering News*, September/October 2002

References

Almgren, Robert F. (2002) Financial derivatives and partial differential equations. *American Mathematical Monthly*, **109**:1–12.

Andersen, L. and M. Broadie (2001) A primal–dual simulation algorithm for pricing multi-dimensional American options. Working paper, University of Columbia, New York.

Bass, Thomas A. (1999) *The Predictors*. London: Penguin.

Baxter, Martin and Andrew Rennie (1996) *Financial Calculus: An Introduction to Derivative Pricing*. Cambridge: Cambridge University Press.

Björk, Thomas (1998) *Arbitrage Theory in Continuous Time*. Oxford: Oxford University Press.

Black, Fischer (1989) How to use the holes in Black and Scholes. *Journal of Applied Corporate Finance*, **1**:4, Winter:67–73.

Black, F. and M. Scholes (1973) The pricing of options and corporate liabilities. *Journal of Political Economy*, **81**:637–659.

Boyle, P. P. (1977) Options: A Monte Carlo approach. *Journal of Financial Economics*, **4**:323–338.

Boyle, Phelim, Mark Broadie and Paul Glasserman (1997) Monte Carlo methods for security pricing. *Journal of Economic Dynamics and Control*, **21**:1267–1321.

Broadie, Mark and Paul Glasserman (1998) Introduction to Chapter III: Volatility and correlation. In Mark Broadie and Paul Glasserman, eds, *Hedging with Trees*. London: Risk Books.

Brzeźniak, Zdislaw and Tomasz Zastawniak (1999) *Basic Stochastic Processes*. Berlin: Springer.

Capiński, Marek and Ekkehard Kopp (1999) *Measure, Integral and Probability*. Berlin: Springer.

Clewlow, Les and Chris Strickland (1998) *Implementing Derivative Models*. Chichester: Wiley.

Cochrane, John H. (2001) *Asset Pricing*. Princeton, NJ: Princeton University Press.

Corless, Robert M. (2002) *Essential Maple 7*. Berlin: Springer.

Cox, John C., Stephen A. Ross, and Mark Rubinstein (1979) Option pricing: a simplified approach. *Journal of Financial Economics*, **7**:229–263.

Cyganowski, Sasha, Lars Grüne and Peter E. Kloeden (2002) MAPLE for jump–diffusion stochastic differential equations in finance. In S. S. Nielsen, ed., *Programming Languages and Systems in Computational Economics and Finance*, Boston, MA: Kluwer, pp. 441–460.

Dalton, John (ed.) (2001) *How the Stock Market Works*, 3rd edn. Englewood Cliffs, NJ: Prentice Hall Press.

Denney, Mark and Steven Gaines (2000) *Chance in Biology*, Princeton, NJ: Princeton University Press.

Duffie, Darrell (2001) *Dynamic Asset Pricing Theory*, 3rd edn. Princeton, NJ: Princeton University Press.

Elder, Alexander (2002) *Come into My Trading Room: a Complete Guide to Trading*. Chichester: Wiley.

Estep, Donald (2002) *Practical Analysis in One Variable*. Berlin: Springer.

Farmer, J. Doyne (1999) Physicists attempt to scale the ivory towers of finance. *Computing in Science and Engineering*, November:26–39.

Forsyth, P. A. and K. R. Vetzal (2002) Quadratic convergence for valuing American options using a penalty method. *SIAM Journal on Scientific Computing*, **23**:2095–2122.

Fröberg, Carl-Erik (1985) *Numerical Mathematics*. Menlo Park, CA: Benjamin/Cummings.

Fu, M., S. Laprise, D. Madan, Y. Su. and R. Wu (2001) Pricing American options: a comparison of Monte Carlo simulation approaches. *Journal of Computational Finance*, **4**:39–88.

Gard, Thomas C. (1988) *Introduction to Stochastic Differential Equations*. New York: Marcel Dekker.

Goodman, Jonathan and Daniel N. Ostrov (2002) On the early exercise boundary of the American put option. *SIAM Journal on Applied Mathematics*, **62**:1823–1835.

Green, T. Clifton and Stephen Figlewski (1999) Market risk and model risk for a financial institution writing options. *Journal of Finance*, **53**:1465–1499.

Grimmett, Geoffrey and David Stirzaker (2001) *Probability and Random Processes*, Oxford: Oxford University Press.

Grimmett, Geoffrey and Dominic Welsh (1986) *Probability. An Introduction*. Oxford: Oxford University Press.

Grinstead, Charles M. and J. Laurie Snell (1997) *Introduction to Probability*. Providence, RI: American Mathematical Society.

Hammersley, J. M. and D. C. Handscombe (1964) *Monte Carlo Methods*. London: Methuen.

Heath, Michael T. (2002) *Scientific Computing: An Introductory Survey*, 3rd edn. New York: McGraw-Hill.

Higham, Desmond J. (2001) An algorithmic introduction to numerical simulation of stochastic differential equations. *SIAM Review*, **43**:525–546.

Higham, Desmond J. (2002) Nine ways to implement the binomial method for option valuation in MATLAB. *SIAM Review*, **44**:661–677.

Higham, Desmond J. and Nicholas J. Higham (2000) *MATLAB Guide*. Philadelphia, PA: SIAM.

Higham, Desmond J. and Peter E. Kloeden (2002) MAPLE and MATLAB for stochastic differential equations in finance. In S. S. Nielsen, ed., *Programming Languages and Systems in Computational Economics and Finance*, pp. 233–269. Boston, MA: Kluwer.

Hull, John C. (2000) *Options, Futures, and Other Derivatives*, 4th edn. Englewood Cliffs, NJ: Prentice-Hall.

Hull, J. C. and A. White (1987) The pricing of options on assets with stochastic volatilities. *Journal of Finance*, **42**:281–300.

Isaac, Richard (1995) *The Pleasures of Probability*. Berlin: Springer.

Iserles, Arieh (1996) *A First Course in the Numerical Analysis of Differential Equations*. Cambridge: Cambridge University Press.

Jäckel, Peter (2002) *Monte Carlo Methods in Finance*. Chichester: Wiley.

Johnson, Philip McBride (1999) *Derivatives, a Manager's Guide to the World's Most Powerful Financial Instruments*. Columbus, OH: McGraw-Hill.

Karatzas, I. and S. Shreve (1998) *Methods of Mathematical Finance*. New York: Springer.

Kelley, C. T. (1995) *Iterative Methods for Linear and Nonlinear Equations*. Philadelphia, PA: SIAM.

Kloeden, Peter E. and Eckhard Platen (1992) *Numerical Solution of Stochastic Differential Equations*. Berlin: Springer (corrected 1999).

Kritzman, Mark. P. (2000) *Puzzles of Finance: Six Practical Problems and Their Remarkable Solutions*. Chichester: Wiley.

Kuske, R. and J. B. Keller (1998) Optimal exercise boundary for an American put option. *Applied Mathematical Finance*, **5**:107–116.

Kwok, Y. K. (1998) *Mathematical Models of Financial Derivatives*. Berlin: Springer.

Leisen, Dietmar P. J. (1998) Pricing the American put: a detailed convergence analysis for binomial methods. *Journal of Economic Dynamics and Control*, **22**:1419–1444.

Leisen, Dietmar and Matthias Reimer (1996) Binomial models for option valuation – examining and improving convergence. *Applied Mathematical Finance*, **3**:319–346.

Lewis, Michael (1989) *Liar's Poker*. London: Hodder & Stoughton.

Lo, Andrew W. and Craig MacKinlay (1999) *A Non-Random Walk Down Wall Street*. Princeton, NJ: Princeton University Press.

Longstaff, F. A. and E. S. Schwartz (2001) Valuing American options by simulation: a simple least-squares approach. *Review of Financial Studies*, **14**:113–147.

Lowenstein, Roger (2001) *When Genius Failed*. London: Fourth Estate.

Madan, Dilip B. (2001) On the modelling of option prices. *Quantitative Finance*, 1.

Madras, Neal (2002) *Lectures on Monte Carlo Methods*. Providence, RI: American Mathematical Society.

Malkiel, Burton G. (1990) *A Random Walk down Wall Street*. New York: Norton.

Manaster, S. and G. Koehler (1982) The calculation of implied variances from the Black–Scholes model: a note. *Journal of Finance*, **38**:227–230.

Mantegna, Rosario N. and H. Eugene Stanley (2000) *An Introduction to Econophysics: Correlations and Complexity in Finance*. Cambridge: Cambridge University Press.

Mao, Xuerong (1997) *Stochastic Differential Equations and Applications*. Chichester: Horwood.

Merton, R. C. (1973) Theory of rational option pricing. *Bell Journal of Economics and Management Science*, **4**:141–183.

Mitchell, A. R. and D. F. Griffiths (1980) *The Finite Difference Method in Partial Differential Equations*. Chichester: Wiley.

Morgan, Byron J. T. (2000) *Applied Stochastic Modelling*. London: Arnold.

Morton, K. W. and D. F. Mayers (1994) *Numerical Solution of Partial Differential Equations*. Cambridge: Cambridge University Press.

Nahin, Paul J. (2000) *Duelling Idiots and Other Probability Puzzlers*. Princeton, NJ: Princeton University Press.

Nielsen, Lars Tyge (1999) *Pricing and Hedging of Derivative Securities*. Oxford: Oxford University Press.

Øksendal, Bernt (1998) *Stochastic Differential Equations*, 5th edn. Berlin: Springer.

Ortega, J. M. and W. C. Rheinboldt (1970) *Iterative Solution of Nonlinear Equations in Several Variables*. PA: re-published by Society for Industrial and Applied Mathematics, Philadelphia, in 2000.

Poon, S.-H. and C. Granger (2003) Forecasting volatility in financial markets. *Journal of Economic Literature*, to appear.

Rebonato, Riccardo (1999) *Volatility and Correlation: In the Pricing of Equity, FX and Interest-Rate Options*. Chichester: Wiley.

Ripley, B. D. (1997) *Stochastic Simulation*. Chichester: Wiley.

Rogers, L. C. G. (2002) Monte Carlo valuation of American options. *Mathematical Finance*, **12**:271–286.

Rogers, L. C. G. and E. J. Stapleton (1998) Fast accurate binomial pricing of options. *Finance and Stochastics*, **2**:3–17.

Rogers, L. C. G. and O. Zane (1999) Saddle-point approximations to option prices. *Annals of Applied Probability*, **9**:493–503.

Rosenthal, Jeffrey S. (2000) *A First Look at Rigorous Probability Theory*. Singapore: World Scientific.

Seydel, Rudiger (2002) *Tools for Computational Finance*. Berlin: Springer.

Smith, A. L. H. (1986) *Trading Financial Options*. London: Butterworths.

Strikwerda, J. C. (1989) *Finite Difference Schemes and Partial Differential Equations*. Belnout, CA: Wadsworth and Brooks/Cole.

Taleb, Nassim (1997) *Dynamic Hedging: Managing Vanilla and Exotic Options*. Chichester: Wiley.

Walker, Joseph A. (1991) *How the Options Markets Work*. Englewood Cliffs, NJ: Prentice-Hall Press.

Walsh, John B. (2003) The rate of convergence of the binomial tree scheme. *Finance and Stochastics*, to appear.

Wilmott, Paul (1998) *Derivatives*. Chichester: Wiley.

Wilmott, Paul, Sam Howison and Jeff Dewynne (1995) *The Mathematics of Financial Derivatives*. Cambridge: Cambridge University Press.

Index